CAD, 3D Modeling, Engineering Analysis, and Prototype Experimentation

Jeremy Zheng Li

CAD, 3D Modeling, Engineering Analysis, and Prototype Experimentation

Industrial and Research Applications

Springer

Jeremy Zheng Li
University of Bridgeport
Bridgeport, CT, USA

ISBN 978-3-319-05920-4 ISBN 978-3-319-05921-1 (eBook)
DOI 10.1007/978-3-319-05921-1
Springer Cham Heidelberg New York Dordrecht London

Library of Congress Control Number: 2014944530

Printed on acid-free paper

Springer is part of Springer Science+Business Media (www.springer.com)

Preface

Computer-aided design (CAD), 3D modeling, and engineering analysis can be efficiently applied in many research and industrial fields including aerospace, defense, automobile, consumer product, and many other product development. These efficient research and engineering tools apply computer-assisted technology to perform 3D modeling on different products, support geometrical design, make structural analysis, assist optimal product design, create graphic and engineering drawings, and generate production documents. This technology helps scientists and technical professionals efficiently import basic geometrical inputs and design information to accelerate the engineering design process, with well-controlled design documents, to support production and manufacturing processes. Currently these research and engineering tools have been playing more and more important roles in different businesses and enterprises due to their financial and technical importance in business, industrial, engineering, and manufacturing applications. The computer-aided modeling and analysis allow more sophisticated, flexible, reliable, and cost-effective manufacturing control. Automation and automated production system are to use control system to reduce human labor intervention during manufacturing processes and put strong impact on industries. Automation and automated system design not only raise the production rate but also control the product quality. It can effectively keep consistent product quality, reduce production lead time, ease material handling, maintain optimal work flow, and meet the product requirement by controlling the flexible and convertible manufacturing/production processes. Computer-aided modeling and engineering design can quickly simulate and model the automated production systems and reduce product development life cycles. Computer-aided engineering solution can improve and optimize the industrial integral processes in design, development, engineering analysis, and product manufacturing. Also the present and future economic globalization requires cost-effective manufacturing via highly industrial automation, efficient design tooling, and better production control. This book describes the technology, types, and general applications of these research and engineering tools through conceptual analysis and real case study in computer-aided design, 3D modeling, and engineering analysis. Some new product systems, developed by author, are introduced to help readers understand how to design and develop new product systems by using computer-aided design, engineering analysis, and prototype experiment. The case studies include design and development of green/sustainable energy systems (solar still, solar panel, and wind power energy), biomedical and surgical instruments, energy-saving cooling system, automated and high-speed assembly system (highly viscous liquid filling and chemical gas charging), robotic system for industrial/automated manufacturing, magnetic sealing system, and high-speed packaging machinery system. Multiple engineering case studies in this book aim at the introduction, study, and analysis by using computer-aided modeling and engineering analysis for industrial and engineering applications. All these newly developed product systems have also been verified by prototyping and testing to validate the functionality of these new systems. Both computer-aided analysis and experimental methodologies introduced in this book show close results that positively show the feasibility and credibility of analytic and experimental methodologies introduced in this book.

Bridgeport, CT, USA Jeremy Zheng Li

Contents

1 Introduction

3D modeling can perform mathematic and geometric analysis on 3D object surfaces via CAD software (Gupta et al. 2010). 3D models can be expressed as 2D images via process of 3D rendering and used in computer-aided simulation to study physical phenomena (Kim and Kim 2011). 3D models can also be geometrically created by 3D printing process. 3D modeling technology allows efficient modeling processes including curve-controlled modeling that can simulate the motion of 3D objects instead of only static geometry (Senthil et al. 2013). 3D computer graphics software can assist 3D modeling processes to create 3D geometrical models. 3D models can represent 3D objects by collecting points connected by different geometric entities including lines, triangles, squares, rectangles, curved surfaces, and irregular geometries in three-dimensional space (Lee et al. 2010). 3D modeling is widely utilized in many different areas, such as 3D graphics design, product development, and computer games (Sun et al. 2005). 3D modeling processes include solid modeling that defines object by volume and shell/boundary modeling that determines object by defining surfaces and boundary (Sipiran and Bustos 2010). 3D modeling can transform all object points, such as internal points and points on circumference surfaces, into polygon elements representing the sphere and volume for model rendering (Ouertani et al. 2011). Triangular modeling meshing is widely used since the meshes can be easily rendered. Polygon meshing element is another modeling method but it is not very popular since the tessellation processing is not provided in the transition to achieve rendering surfaces (Li et al. 2012). 3D polygonal modeling is one of the most popular modeling methods due to its accurate, flexible, and quick meshing process. In 3D polygonal modeling, 3D points are linked via many tiny line elements to generate polygonal meshes (Harik et al. 2008). 3D curved modeling is another common method, in which all the object surfaces are specified by curves that are manipulated by the weight-controlled points in 3D space. The curves will move close to the points when weight of these 3D points is increased for more accurate modeling process (Reich and Paz 2008). Compared to 2D modeling methodology, 3D modeling can change and animate parts with (1) quick object rendering, (2) easier object

J. Zheng Li, *CAD, 3D Modeling, Engineering Analysis, and Prototype Experimentation*, DOI 10.1007/978-3-319-05921-1_1

rendering, and (3) more accurate rendering (Tian et al. 2009). 3D modeling has been applied in different businesses and industries including movie filming, consumer product design, industrial design, cartoon animation, video gaming, architecture design, and engineering research (Walthall et al. 2011). CAD software can be used to assist 3D modeling for product design and development.

Computer-aided design (CAD) is to apply computer systems to assist the engineering process for creating, modifying, analyzing, and optimizing the product design (Stefano et al. 2013). CAD software is used to accelerate design process, improve design quality, ease technical communication via engineering documentation, and build database for production (Veltkamp et al. 2011). CAD results can be output in electronic files for printing, manufacturing process, production operation, etc. The CAD systems can be applied in different product designs including electronic, civil, mechanical, and automated systems (Pessoa et al. 2012). CAD is an efficient engineering design tool that has been widely used in different applications including designs of car, ship, aircraft, industrial products, and architecture. (Starly et al. 2005). CAD can also be extensively applied to generate computer-aided animation for filming, commercial advertising, and product manuals (Kosmadoudia et al. 2013). Current CAD software packages provide 2D drafting and 3D solid modeling. CAD can allow three-dimensional object rotation, view designed object from different angles, and check full geometrical features from inside and outside of desired objects (Piatt et al. 2006). CAD can be applied for building conceptual design and product layout, defining production methods via structural analysis of product assembly, and detailing engineering 3D models/2D manufacturing drawings (Sung et al. 2011). CAD systems were originally developed with computer languages including Algol and Fortran but CAD technology has been significantly changed due to development of object-oriented programming (Vincent et al. 2013). Modern CAD systems have been developed using interaction of graphical user interface with object geometry and boundary envelop to control relationships among different object geometries in complex sketches, part models, and product assemblies (López-Sastre et al. 2013). Currently CAD systems can work with most platforms such as Windows, UNIX, Mac OS X, and Linux. Today there are many different CAD systems applied in business, research, engineering, and industry including Pro/Engineer, SolidWorks, CATIA, Solid Edge, Inventor, Unigraphics, CADDS, and AutoCAD (Rocca 2012). Computer-aided design and engineering analysis have been applied to create 3D product features, specify the material information in mechanical and thermal properties, define geometrical shape, determine part dimension, perform manufacturing tolerance control, and analyze the system functionality and structure of product systems (Adán et al. 2012; Chae et al. 2011). CAD technology significantly reduces drafting time and efficiently helps professionals in product design and development (Chaouch and Verroust-Blondet 2009). Current CAD software packages provide efficient ways to control product design in 3D space, make engineering drawings quickly, and allow users easily review product design in different views to accelerate the design process (Claes et al. 2011). Compared to the manual drafting design, CAD technology can significantly shorten the design time, improve design quality, and optimize complex geometric design (Adams and Yang 2004). CAD technology can be applied to assist geometric dimensioning and tolerancing (GD&T) control, create conceptual design, make assembly layout, and perform kinematic and dynamic analysis (Goel et al. 2012). 3D geometrical parameters and boundary conditions can be used to specify the product dimensions, shape, and solid elements (Bertoni and Chirumalla 2011). The computer-aided engineering analysis (CAE) can be used with CAD to determine the structural strength of products including tensile, yield, principal, and shear strength (Ding, et al. 2009). CAD system can also be used to perform graphic simulations for preparing different enterprise documents, such as project of environmental protection in which the CAD-assisted constructions can be superimposed into existing environmental graphic piles to determine what effects will be caused to the environment if targeted constructions are being built (Catalano et al. 2011). Computer-aided design of automated system brings cost-effective processes to control complex manufacturing systems and production in industry (Fuge et al. 2012).

In this book, the CAD software of Pro/Engineer is utilized for 3D solid modeling/product design and Autodesk simulation software is used for engineering simulation/structural analysis.

1.1 Solar Energy System for Water Distillation

People can have daily clean and pure drinking water easily since getting clean water is simply opening the faucet. However, in many underdeveloped countries or in some extreme disaster-related situations, it is difficult to get clean and pure water (Anjaneyulu et al. 2012). Solar distilling process is a way of changing impure water into clean water. Based on report from the World Health Organization, about 1.1 billion people over the world are not able to find safe drinking water. Among them, about 2.1 million people die each year due to drinking of contaminated water (Badran et al. 2005). The solar distilling process is a method of distilling water by using the heat from the sun to generate moisture evaporation from humid environment and applying air to cool the condenser to produce filtrated water. Distillation process is one of the methods to control water purification (Jabbar et al. 2009). Sunlight is one of multiple heat energies that can be applied to perform water distillation process. In solar water distillation process, there is no fuel cost but requires associated costly distilling equipment (Manikandan et al. 2013). Although the solar distilling drinking water costs several times that of water supplied from city utilities, it is still less expensive than the bottle water in outside store due to its energy-wise distilling process (Lattemann and Höpner 2008). In case the local residents are worried about purification quality or concerned about the purified addictives added to the local city water, solar distilling of tap water will be a safe and energy-saving process (Chakraborty et al. 2004). Since the energy cost is continuously increased and the pressure of more human population is constantly exerted on current available freshwater, the solar desalination of seawater has its energy-efficient and cost-economic advantages (Jabbar et al. 2009; Li 2011c). In solar still unit, the impure water is gathered around the outside surface of collector and evaporated by sunlight that is absorbed through clear plastic panel. When pure water vapor passes the condenser, it will get cooled and condensed on the cold surface. The filtrated water droplet will drip down by its gravity to the pure water collector at lower chamber in solar still unit. This distilling process takes away the impurities including heavy metals and microbiological organisms from environmental water (Tiwari and Tiwari 2007). The solar still system can also be applied in the places where rainwater, well water, or city water is not available. In case of power outage during severe weather conditions, such as hurricane season, the solar distillation system can supply an alternative clean water resource. The basic basin-type solar still unit mainly consists of some stones, transparent plastic or glass panel, condenser, and collector to store condensed pre-water (Yang et al. 2011). As the sun heats and evaporates the moisture, water vapor moves to condenser where the vapor gets cold and condensed to form water droplet which will drop down to pure water collector at the bottom of solar still unit. Other solar distill systems, such as wick solar still, can distil the salt water. In wick solar still system, salted water input in from the top gets evaporated after heated by the sunlight through transparent plastic or glass panel (Alloway 2000). The vapor starts condensation at the underside of plastic panel and drips to the bottom collector. The purity of distilled water stored in the bottom collector relies on how much salt can be separated from the salt water in solar distillation unit. If more wicks are constructed in the solar still, more heat can be transmitted to the salt water which makes more distilled water product. A plastic fine grid thin plate can be installed in order to capture more brine from salt water before it goes down to the container. This will provide longer time to heat up impure water and separate the brine from salt water. The wick-type solar still should be equipped with good seal in order to prevent vapor from escaping to the outside environment. Some wicks should be darkened in order to absorb more heat to increase distilled water productivity (Jabbar et al. 2009). There are several other different types of solar sill designs including the single-basin distillation unit that consists of a basin equipped with a tilted thin glass or plastic plate to hold impure water. The dark basins can function better to capture the sunlight energy. The solar distillation units equipped with glass usually show durable function and longer life but the units equipped with plastic sheet are of lower cost and have easy installations. The tilted thin glass or plastic plate permits the water to easily drain out of the solar distillation units into the collector through a tube (Anjaneyulu et al. 2012).

1.2 Wind Power Turbine System

The wind power is a process in which the wind turbine converts wind energy into mechanical (kinetic) energy (Ogbonnaya 2011; Passon et al. 2007). The mechanical energy can be applied to generate the electricity in wind power plant system, or employed to operate machinery or pumping water in windmill or wind pump system (Agarwal and Manuel 2007; Simhauser 2010; Saravanamuttoo et al. 2009).

Wind power density which is related to the wind velocity and air density can be used to calculate the mean annual power generated in each square meter of turbine sweeping sectional area and the density changes with different heights (Bir and Jonkman 2007; Kim et al. 2011; Li 2013; Vallee et al. 2009). In the real wind power turbine, it is not possible to capture total wind power since some acquired air will exit the turbine system. The ratio of inlet and outlet wind velocity should be considered in the wind turbine system design and the maximum efficiency of gained wind power by current turbine is around 60 % (Carey 2010; Li 2012f; Singh and Nestmann 2011). The power delivered by wind turbine system will be reduced due to the losses in gear train, converter, rotor blade, and generator (Fulton et al. 2006). The turbines are normally placed at upwind location of structural tower and turbine rotor blades are constructed in strong stiffness to keep the blades from being bended into structural tower due to strong gusty wind (Christodoulou et al. 2011; Li 2012f; Ogbonnaya et al. 2010). Wind turbine systems have been designed to capture the wind energy in a specific place and aerodynamic analysis can be employed to verify the proper height of structural tower, to decide the feasible control systems, and to determine the rotor blade geometry and numbers (Komandur and Sunder 2008; MacLeod and Jastremski 2010; Silva et al. 2011).

1.3 Solar Panel Tracking System

The global warming demands and requests the alternate energy resources from green and renewable energy sources including solar power energy. The solar panel tracking systems are the device that orients solar panel following movement of the sun (Bhandari and Stadler 2009; Munilla 2013). Solar panel can be photovoltaic and reflective panels or some optical related devices. In photovoltaic flat panel system, tracking mechanism is applied to reduce the incidental angle between input sunlight and solar panel to increase the incoming energy received from the sun. In concentrated solar photovoltaic system, the tracking mechanism is employed to orient optical device towards the sun to receive maximum direct sunlight energy (Brinkworth and Sandberg 2006; Li 2013b). The effective sunlight-receiving area in solar panel system changes with the cosine of angular deviation between panel direction and the sun (Hoke and Komor 2012). Since sunlight has two components in which around 90 % of solar energy is contained in direct sunlight and rest energy is contained in diffusive sunlight, the sun requires to be visible as much as possible; otherwise more direct sunlight energy will be proportionately reduced in cloudy sky (Darling et al. 2011). The tracking system with accuracies of $\pm 4.5°$ can catch more than 98.8 % of the energy from direct sunlight and also 100 % of the diffusive sunlight (Laird 2011; Mendonça and Jacobs 2009). Although the sun moves 360° from east to west each day, the approximate visible portion of the sun is around 180° (average half day time). If a solar panel in horizontal location does not rotate from east (dawn) to west (sunset), only sunlight that travels about 80° could be caught and rest of the sunlight energy in the early morning and late afternoon will be lost (Timilsina et al. 2012). The solar panel orientation from east to west can help to maximize the capture of daily sunlight energy. Single-axial tracking system that has one degree of freedom with angular rotation around one axis can be applied to this solar system (Wang et al. 2011). There are several types of single-axial tracking systems including horizontal, vertical, inclined, and polar aimed single-axial tracking systems. In horizontal single-axial tracking system, the orientating axis is installed horizontally related to the ground and setup of multiple solar panel groups is simple since all rotating axes of all groups can be maintained parallel to each other. In vertical single-axial tracking system, the orientating axis is mounted vertically related to the ground. This tracking system rotates solar panel from east to west during the day and it works more efficiently than horizontal axial solar tracking system at higher elevation (Myers et al. 2010). The setup of multiple groups in vertical axial tracking systems should consider reducing the shade to minimize unexpected sunlight energy losses. In tilted single-axial tracking system, it has two rotating axes with which the solar panel can rotate around horizontal axis and vertical axis individually. The tilted angle system can be adjusted to decrease the wind load pressure (Simhauser 2010). The multiple group setups are needed to reduce the shade to minimize the sunlight energy losses. Although the panels in group setups can be adjusted without shade when perpendicular to the rotating axis, the setups that parallel to their rotating axes are complicated and will be limited by the panel tilted angles and its elevations. The polar aimed single-axial tracking system equips a telescope-guided unit and tilted single axis is adjusted to aim at the polar star. In this tracking system, the solar system's tilted axial angle equals to its site latitude that keeps good alignment between rotating axis of tracking system and orientating axis of the earth. The sun travels 48° between the north and south as well in a full year and only sunlight that travels about 24° could be caught (Branker et al. 2011). The tracking system with orientation from the east to the west (daily movement) and from the north to the south (seasonal movement) called double-axial solar tracking system should be used to minimize the lost sunlight energy. The double-axial tracking system that has two degrees of freedom with angular rotation around two axes can also be employed to the solar system. In this system, two axes are normally perpendicular to each other (Ramadhan and Naseeb 2011). The primary axis is the rotating axis related to the ground and secondary axis is normal to the primary axis. Several major types of double-axial tracking systems include tip-tilted double-axial tracker and azimuth-altitudinal double-axial tracker (Falconett and Nagasaka 2010). The rotation of solar panel related to the tracking system is important to the solar function and double-axial tracker permits better receiving of solar energy due to its capability of tracking the sun in vertical and horizontal directions. Some factors should be considered while selecting the types of tracking systems, such as environment condition, local latitude, on-site weather, electrical price, and installation dimensions. The tracking systems using motor drivers and gearing unit to orientate the solar panels are directed by controller to track the sun traveling directions (Ferrey 2006; Ramadhan and Naseeb 2011).

1.4 Energy-Saving Cooling System

The cooling process is to move heat from one place to another place and heat transfers are driven not only by mechanical energy but also by the energies from electricity, heat, and magnetism (Bagarella et al. 2013; Christian and Hermes 2013; Lucas and Koehler 2012). There are many applications using cooling systems including cryogenics, commercial freezers, residential refrigerators, and air conditioners (Agrawal and Karimi 2012; Barbosa and Sigwalt 2012; Li 2009d). The current widely used applications of cooling systems are for industrial cooling processes in manufacturing and production, air conditioning for residential and commercial buildings, medical treatments, surgical operations, climate-controlled food conservation, and many others (Derking et al. 2012; Engelbrecht et al. 2012; Oró et al. 2012a, b). In industrial and production processes, the cooling systems can be applied for gas liquefaction, air purification, oil refinery, and metallic material temper treatment (Khan et al. 2012; Li 2012; Mumanachit et al. 2012). There are several cooling processes including cyclic cooling (i.e., vapor cycle and gas cycle), magnetic cooling, and thermoelectric cooling (Ally et al. 2012; Sanaye and Asgari 2013). In cyclic cooling process, heat is taken away from lower temperature source and released to higher temperature source that is driven by external energy work. The refrigerant absorbs and releases the heat as it circulates in a cooling system (Bhanja and Kundu 2011; Shamsoddini and Khorasani 2012). The cyclic cooling processes include vapor cooling cycle and gas cooling cycle (Egolf et al. 2012; Zhu et al. 2013). In vapor cooling cycle, vapor is compressed with no change of entropy and leaves compressor at higher temperature but pressure is still below the vapor pressure at that associated temperature. The vapor starts condensing into the liquid phase after passing the condenser and the evaporation will be generated when liquidized refrigerant flows across the expansive valve to absorb the heat through evaporator unit to cool external environment (Colombo et al. 2012; Cuevas et al. 2012). In gas cooling cycle, the gas phase does not change during compression and expansion. The cooling capacity equals the product of gaseous specific heat and temperature rise in lower temperature source (Abed et al. 2013; Thomas et al. 2012). The gas cooling cycle is widely applied as cooling system in gas turbine-driven fight jet or airplane. In thermoelectric cooling system, the heat flow between two different media contacts is driven by Peltier effect that is widely applied in mobile cooling unit for temperature control of instruments and electronic devices (Coşkun et al. 2012; Šarevski and Šarevski 2012). In magnetic cooling system, the refrigerating media is usually a paramagnetic salt and active magnetic dipoles are from the electron shells of paramagnetic atoms (Ayou et al. 2012; Faúndez et al. 2013). The multiple magnetic dipoles in cooling media are driven to be aligned under higher magnetic field and degrees of freedom in cooling media are being put into lower entropy environment (Aprea et al. 2013; Qureshi and Zubair 2013). The thermal sink then receives heat removed from cooling media due to its entropic loss. The contacts between cooling media and thermal sink is then disconnected and the unit is insulated causing switch-off in magnetic field that raises thermal capacity of cooling media and reducing its temperature below thermal sink temperature (Gheisari et al. 2012; Kagawa et al. 2013).

The coefficient of performance (COP) of a cooling system is a critical parameter in judging the systematic efficiency of a cooling system (Kitanovski et al. 2012; Piacentino and Talamo 2013). COP is designated as the ratio of cooling capacity to external energy input (Kumlutaş et al. 2012). The performance factor (PF) of a cooling system is another important parameter in examining the systematic efficiency of a cooling system. PF is determined by the ratio of energy input to cooling capacity (Ekren et al. 2011; Góral and Kluza 2012). The newly developed cooling system introduced in this book focused on energy saving and manufacturing cost reduction due to its simplified design.

1.5 Automated and High-Speed Manufacturing Systems

Automated machinery system is to apply various control systems to control and operate the machinery with minimized or decreased human intervention (Bao et al. 2013; Janchiv et al. 2013). The major advantages of automated machinery system include reducing labor, saving energy, optimizing material usage, improving quality, keeping accuracy, and maintaining precision (Andrikopoulos et al. 2013; Bay et al. 2008; Jeon et al. 2013). Automated machinery systems can be supported by many different ways including computers, pneumatics, hydraulics, mechanics, and electronics (Baniardalani and Askari 2013; Kim et al. 2013). Two popular controls applied to automated machinery system are feedback control and sequential control (Koo et al. 2013; Li 2011b; Liu et al. 2013). The feedback control includes continuous measurement by applying optical or proximity sensors and computes adjustments to maintain the allowable range for measured variables (Bang et al. 2013; Beebe 2009; Liaquat and Malik 2013). For example, when using feedback control for air-heating system, the sensor detects the heated air temperature and the signal is being continuously fed back to the systematic controller to compare with targeted variable setting (Berretti et al. 2012; Che et al. 2013; Lini et al. 2013). The temperature difference is calculated in the controller and the signal will be sent to the heater after temperature adjustment has been determined (Nikolakopoulos and Alexis 2013; Pai 2013; Tombari et al. 2010a, b). The sequential control involves executing programmed logic sequence in discrete operations. The relay logic is one of the forms in sequential control and electrical relays control electric contacts that can make either connection or disconnection between the electronic devices (Berretti et al. 2012; Cho et al. 2013; Park et al. 2013). For example, when applying sequential control for elevator, relay logic has been created while engaging or disengaging the electrical contacts to start or stop electric motor to operate the elevator (Bouazza and Ouali 2013; Li 2012; Lini et al. 2013). Machine tools can be automatically operated by numerical control (NC). Automated machinery systems can be applied in manufacturing and assembly processes, such as petroleum refining, power generation, chemical production, plastic molding and injection, steel making, automobile assembly, food processing, automated welding, and many other industrial applications (Choi et al. 2013; Sami and Patton 2013; Tuan et al. 2013). The major advantages using automated machinery systems include higher productivity, better quality, more consistency, less human involvement, and reduced labor cost (Faltemier et al. 2008; Ullah et al. 2013). The automated system is preferably used, where the cyclic time has to be reduced, higher accuracy must be maintained, working environment is hazardous or dangerous to the human, and job task is far beyond the operator's capability, to keep strong economic competitiveness of enterprises in the challenging market (Lee et al. 2013; Seifabadi et al. 2013). Automated machinery system can meet the growing demand for flexibility and convertibility in production processes. The manufacturing industries are continuously demanding the capability to flexibly change from making one product to another new product with no need to completely set up a new production line (Han et al. 2013; Li 2011b; Seok et al. 2013). The numerical control (NC) has been currently applied to control automated production systems in many companies quickly expanding the automated applications and widely benefiting the human activities (Christophe et al. 2010; Hu et al. 2013; Van et al. 2013). Computer-aided control technology can be employed as the basis to create complicated industrial systems through implementing mathematical and business's functioning models, such as computer-aided design (CAD), computer-aided engineering analysis (CAE), and computer-aided manufacturing (CAM) (Ding et al. 2013; Han et al. 2013; Wang and Yang 2013). The integration of information technology (IT) with industrial machines and manufacturing processes can significantly support the control system design and development, such as programmable logic control (PLC) system (Jeon et al. 2013; Ji et al. 2013; Lamooki 2013). The PLC system can be normally used to control the operation sequence between input sensors and output actuators. The automated control systems can be widely applied in different industries including producing aerospace, food, mining, automotive, environmental waste process, medicines, agricultural products, chemicals, metal working, product inspection, and machine cutting (Huang et al. 2010; Kong and Tomizuka 2013; Lee et al. 2013). The importance to increase manufacturing rate is to integrate the automated software with machinery drives, production services, and business solutions to make company more competitive (Li et al. 2013; Yang et al. 2013).

1.6 Robotic System for Industrial Applications

Robotic technology is a technical field dealing with engineering design, manufacturing operation, automatic applications, computational control, sensing feedback, and data processing (Howard et al. 2006). The robotic system design usually integrates the technologies of computer science, mechanical engineering, electronic engineering, and manufacturing engineering (Kazerooni 2005; Li et al. 2011). The mathematical expression of a robotic system is used to control algorithms in observing how a functioning process has been handled (Harja et al. 2007).

In history, the robotic systems were sometime being used to simulate human behavior and perform certain tasks. Today, the robotic systems are rapidly developing with continuous advances of automation technology, scientific research, and engineering design (Lerman et al. 2006). The robotic systems can perform different special jobs that are heavy duty, too dirty, very dull, and environmentally hazardous to the humans. Robotic systems are more accurate and reliable than humans so they have been widely applied in production processes including product packaging, manufacturing assembly, material transport, space exploration, surgical procedure, and many others (Pounds et al. 2004; Ratti and Vachtsevanos 2010). The actuators in robotic systems, such as motors and drivers, transfer the stored energy to the different kinds of motions including linear and angular movement (Mian et al. 2010). The mechanism of robotic systems is controlled to perform different functions. The sensors allow robotic systems to obtain data information from measurement and decide the related response to perform different tasks (Kennedy and Desai 2003; Salti et al. 2013). In the robotic operation, sensors receive and provide information of external conditions (i.e., location, temperature, pressure) and situations of robotic system itself (i.e., location of its arm, gripper, joint pin points) (Klopcar et al. 2007; Tapus et al. 2008). All these received data information can be collected, calculated, and executed in systematic controller to instruct the movement of robotic system. The robotic systems can be used for manipulating the products including picking and placing parts in the manufacturing processes (Leibe et al. 2008). The robotic arms, equipped with mechanical or vacuum grippers, can move around in a range of 3D motion by more advanced ways using balanced dynamic algorithm (Mellinger et al. 2010). The robotic motion study focuses on two areas: kinematic motion and dynamic motion. Direct kinematic motion study performs the calculation of gripper location, rotation, speed, and accelerated rate while the associated connecting joint values are given (Rachidi et al. 2013; Shakernia et al. 2002). Inverse kinematics motion study considers the situation in which the connecting joint values are computed while the gripper position values are given (Mian et al. 2008). The kinematics can deal with the collision prevention, singularity elimination, and system redundancy. After getting associated locations, speeds, and accelerations by applying kinematics, the effect of forces on its motion can be determined by employing the dynamics (Shen et al. 2008; Tsagarakis and Caldwell 2003). The direct dynamics focuses on the computation of acceleration in robotic systems if exerted forces are given and inverse dynamics studies the robotic driving forces required to generate the robotic system acceleration (Tapus et al. 2007). All the above information can be applied to adjust and control the algorithms of the robotic systems (Torbjorn et al. 2009).

1.7 Magnetic Sealing System

There are several types of magnetic sealing systems. The magnetic liquid seals are applied in rotating machinery to allow rotation while keeping a tight seal by using ferrofluid, that is contained in place by applying the permanent magnet, as the physical barrier to prevent leaking (González-Jorge et al. 2005; Mitamura et al. 2008). Magnetic liquid rotating seals have almost no requirement of maintenance and very low leakage in the different applications (Shlyk et al. 2004). This type of seal is normally installed in mechanical assemblies with a centralized shaft, support bearings, external housing, and other components. The support bearings offer two critical functions: keeping shaft centralized with even seal gap and holding outside loads. Since the magnetic liquid rotating seal is actually an oil-distributive fluid contained magnetically between rotary shaft and stationary wall inside of housing, the life cycle for this type of seal is increased and frictional torque load is reduced (Cong et al. 2005; González-Jorge et al. 2007). There is no requirement for electrical power since magnet has been permanently charged. Magnetic rotary liquid seal system is designed for many applications for gaseous and vapor sealing but not for pressured fluid sealing due to weakness caused by sharp varied temperature, alternative pressure, high speed, differential loads, and severe environmental changes (Hirsch 2003; Li 2010). Another magnetic seal, also called vacuum seal, relies on the technology of using electromagnetic field to hold metal components together. This magnetic sealing unit is usually applied to seal the nonenvironment-friendly products and keep hazardous materials from leakage (Topal et al. 2003; Zydlo et al. 2005). It has been designed to block any impurities or tiny particles that intended to permeate the seal mechanism and build up an airtight seal between rotary shaft and stationary wall inside the housing (Ochonski 2005; Tušek et al. 2011).

A new type of magnetic sealing system introduced in this book focuses on more reliable functioning sealing system with less manufacturing cost and enlarged life cycles. Currently many rotary machineries use the conventional seals in lip or labyrinth geometrical shapes (Li et al. 2005). However, these conventional seals show significant wear and leakages of gas and liquid through these types of seals were detected (Takashi 2002). The life cycle of these traditional seals is much shorter than that of magnetic seals (Morton and Fruh 2002). The magnetic sealing mechanism can be applied to improve sealing function in different engineering applications (Zhao et al. 2006).

1.8 Automated and High-Speed Packaging Machinery System

The packaging engineering and technology is a technical field ranging from packing system design to final product placement. All the sequential layouts in the production line should be taken into account while designing the package system for any developed end products (Fuge et al. 2012; Han and Lee 2013; Jeon et al. 2013). Packaging production systems are used to reliably package the products to prevent products from damage during delivery and in the storage (Cho et al. 2013; Fusco and Russo 2013). Many innovations in the packaging systems were initially developed for military practices since certain military packaged products must be transported in the most severe environment, harsh distribution, terrible storage, and dangerous conditions (Chu and Chang 2005; Jeong et al. 2013;

Lee et al. 2010). When designing the package engineering systems, it should focus on the industrial and technical aspects in production, marketing, industrial logistics, materials being handled, and final product-related design (Devanathan and Ramani 2010; Harja et al. 2007; Hu et al. 2013). The package being processed must prevent the products from breakdown while keeping the cost-effective and efficient packaging production cycles. The objectives of packaging technique target that the products wrapped in the package should be kept from ruin caused by external compression, physical shock, harsh drop-off, varied temperature, mechanical vibration, and electrostatic discharge (Gerkey and Matarić 2004; Goel et al. 2012; Jiang et al. 2013). The transport packaging needs to follow the standard logistics system in order to meet the protective strength and holding capability of transporting packages. The package design and testing might be performed within the company or from outside packaging engineering firms (Andrikopoulos et al. 2013; Baniardalani and Askari 2013). To pack products by machinery system, the packaging machine selections should consider its packaging capabilities, technical requirements, systematic maintainability, labor involvement, packing reliability, operation safety, machine serviceability, flexibility to be integrated into the production line, layout space, front and running cost, energy efficiency, and ergonomic design for future transport handling (Agarwal and Manuel 2007; Gao et al. 2013; Seifabadi et al. 2013). Automated packaging systems can improve packing quality and increase productivity. There are many types of packaging machineries including systems for cartooning, bottling, accumulation, collection, slitting, sealing, converting, filling, and wrapping (Kim et al. 2013; Salti et al. 2013; Song and Li 2013). Packages are designed from many different types of hard or flexible materials that have folding lines to permit further folding into the package shapes. Some major processes that packaging manufacturing is mainly involved in are extrusion-assisted forming, thermo-assisted forming, molding-assisted forming, and other technology-assisted forming processes (Koo et al. 2013; Regli et al. 2011; Seifabadi et al. 2013). Packaging productions can be designed for high-speed processes including filling, packing, and shipping (Fang et al. 2013; Rezgui et al. 2011). The structural and thermal analysis of packing tools and materials requires to be performed to evaluate the packaging quality and its further improvement (Ratti and Vachtsevanos 2010; Wang et al. 2013). Good packaging system design makes products more tangible to the users, sets a tight customer connection, and efficiently influences marketing decision to engage (Passalis et al. 2011; Yoon 2013). It is very important to understand how to combine global marketing information with customer needs and cultural preference to effectively join current and future compelling global discipline (Stefano et al. 2013; Zhu 2013).

1.9 Biomedical and Surgical Systems

New biomedical and surgical instruments have been designed and developed to perform complex surgery and biomedical treatment by applying the minimally body-invasive approaches and reliable medical instruments (Brown et al. 2007; Cheng et al. 2004; Kavitha and Ramakrishnan 2011). Many different medical treatments and surgical procedures can be facilitated efficiently for gynecology, urology, prostatectomy, general medical treatment, cardiac valve restore, plastic surgery, orthopedics, and neurology by applying good medical instruments (Ballihi et al. 2012; Chambers et al. 2013; Kosta et al. 2012). Doctors and surgeons are required to know the knowledge and learn the skills to correctly use different surgical instruments in sterile and aseptic surgical environment (Fadzil et al. 2011; Kayalvizhi et al. 2013). The goal of newly developed biomedical and surgical instruments is for doctors and surgeons to be able to perform the medical treatments more smoothly and efficiently to ensure that the medical operative procedures are safe and instrument functions are reliable (Gill and Munroe 2012; Goy et al. 2012; Kumar et al. 2012). To carry through a surgical procedure, doctors and surgeons must well control and accurately manipulate the medical instruments to keep full safety and functioning features that are designed for minimum human errors (Hemalatha and Manivannan 2011; Janghel et al. 2012; Li 2012). New biomedical and surgical instruments are designed and developed to improve upon conventional surgical procedures to ease and benefit doctors and surgeons in medical treatments and surgical operations by providing clear visualization, precise control, ergonomic satisfaction, consistent function, safe interaction, and flexible adjustment (Haddad-Mashadrizeh et al. 2013; Kalantzaki et al. 2013; Li 2011). The newly designed and developed biomedical and surgical instruments introduced in this book also aim at eliminating surgiclip drop-off incident, less operational force, more stable driving mechanism, and more robust structural design while manipulating the thick body tissues in medical treatments and surgical operations.

Part I

Energy Systems

Solar Energy System for Water Distillation 2

Some newly designed and developed product systems are introduced in this chapter. It includes the green (sustainable) energy system for water distillation, wind power system, automated and high-speed machinery system, solar panel tracking system, robotic system, biomedical and surgical system, energy-saving cooling system, magnetic sealing system, and high-speed product packaging system. All these new product systems have been designed and developed through computer-aided modeling and simulation. The design methodology has been described for these new product systems to detail the design procedure and assist readers in their future academic endeavor.

2.1 Design of Solar Energy System for Water Distillation

Solar water distillation systems have the potential to make the clean water for safer drink, cooking, and many other commercial usages. The solar still is to apply solar energy to generate the water evaporation and catch the vapor by condensing it at cold surface. The evaporating rate can be controlled through increasing water temperature and enlarging the contact area between water and air. The small individual solar water distillation devices are for residential purpose and larger distilled water system for industrial applications. Since many areas do not have enough water distillation or purification systems, people are sometimes concerned about environmental pollutants in the water supply. The application of solar water distillation system is one of the effective possible solutions. The solar water distillation system has the advantage over regular water filtration systems that still show some potential threats to the people due to the difficulty of removing some inorganic materials. In solar water distillation process, water gets heated and finally becomes steam. The steam is then cooled and condensed into liquid water through heat exchanger and any inorganic materials contained in the water can be

J. Zheng Li, *CAD, 3D Modeling, Engineering Analysis, and Prototype Experimentation*, DOI 10.1007/978-3-319-05921-1_2

Fig. 2.1 Newly developed green (solar) energy distillation system

filtered out. Meanwhile, this solar distillation process does not affect good organic minerals existed in the water. This new green (solar) energy distillation system can also separate the dissolved salts from sea water to make fresh drinking water. This cost-effective solar system is simple and it can be easily installed with less maintenance. It keeps higher water temperature at unit entrance, maintains proper temperature difference between input water and condensing surface, and minimizes the vapor leakage. This new systematic design helps entering water to absorb as much radiation as possible with less heat losses from the floor and walls.

This new green (solar) energy distillation system is displayed in Fig. 2.1.

Water flows in from inlet tube and moves the little ball up to open up the internal channel for entering water which will be collected in the basin thereafter. When target water level is arrived, the large ball travels up and little ball travels down to barricade the inlet tube to keep more water from entering the basin. This solar distillation system will get hot due to the sunrays emitted from the reflector to warm up the absorber. The absorber takes in the sunrays and provides the heat to the impure water entering from inlet tube. Then water begins evaporating and the vapor will be condensed when it arrives at the upper chamber of basin, adhered to the internal wall, and changed to the water droplet because the temperature in upper chamber is lower than that in bottom chamber. The condensed water droplets can travel to the U-type channel in solar energy distillation system to make pure water from any type of impure water including sea water. The purified water from output pipe can be accumulated in the output container.

2.2 Computer-Aided Simulation of Solar Energy System for Water Distillation

Since the sun is extremely far away from the earth, its radiated light can be considered to be almost parallel to the earth surface. The solar energy recorded at the earth surface is around 1,368 W/m^2 (Alloway 2000). Since the earth rotates itself and moves around the sun eccentrically because of its elliptically traveling path, the distance between the sun and earth changes as the earth moves. If ψ is the angle between perpendicular direction to the earth surface and the solar radiated flux flow of N, solar energy caught by panel surface in solar distillation system can be calculated by the equation of $N \times \cos(\psi)$. Angle ψ also changes based on latitude (ζ), declination (λ), hour angle, and zenith angle (Alloway 2000):

$$\lambda = 23.5 \times \sin\left[\frac{360}{365} \times (284 + S)\right] \quad (2.1)$$

In diffusion radiation, tilt factor, the ratio of radiated diffusion flux flow arrived on the tilted surface to the horizontal surface, is represented by the ω_{F} as follows (Gevorkian 2007):

$$\omega_{\mathrm{F}} = \frac{1 + \cos(\eta)}{2} = 0.93 \quad (2.2)$$

The computer-aided modeling and numerical simulation are applied to design and develop this new solar distillation system.

Figure 2.2 shows the computer-aided FEA meshing in this new solar energy distillation system.

Figures 2.3 and 2.4 display the computer-aided simulation of solar radiation intensity in this new solar energy distillation system. The computational simulation from Fig. 2.3 demonstrates that the maximum solar radiation intensity is 2,388.50 W/m^2.

Figure 2.4 exhibits the distilled water output vs. duration time in this new solar distillation system.

Figure 2.4 presents that the distilled water output is proportion to the basin temperature in this solar energy distillation system. If solar basin temperature is raised, the water evaporation will be accelerated to keep more pure water produced from this solar distillation system.

In order to verify if the structural support of solar distillation system is properly designed, the wind load and snow load are being presented in the following mathematical equations (2.3) and (2.4), respectively.

Velocity pressure (q_z) at solar distillation system height z can be calculated by the equation (ASCE 7-05 2005)

$$q_z = 0.00256 \times K_{\mathrm{Z}} \times K_{\mathrm{ZT}} \times K_{\mathrm{D}} \times V^2 \times I \quad (2.3)$$

Fig. 2.2 Computer-aided FEA meshing in new solar distillation system

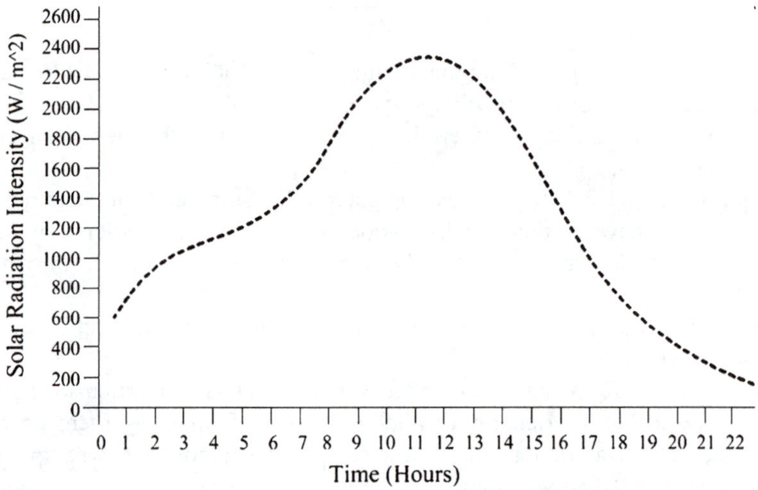

Fig. 2.3 Solar radiation intensity in this solar distillation system vs. duration time

where q_z—effective velocity pressure, K_D—wind directionality factor, K_Z—exposure velocity pressure coefficient, K_{ZT}—topographic factor, V—wind velocity, and I—importance factor.

Snow load (P_F) on solar distillation can be determined by the equation (ASCE 7-05 2005)

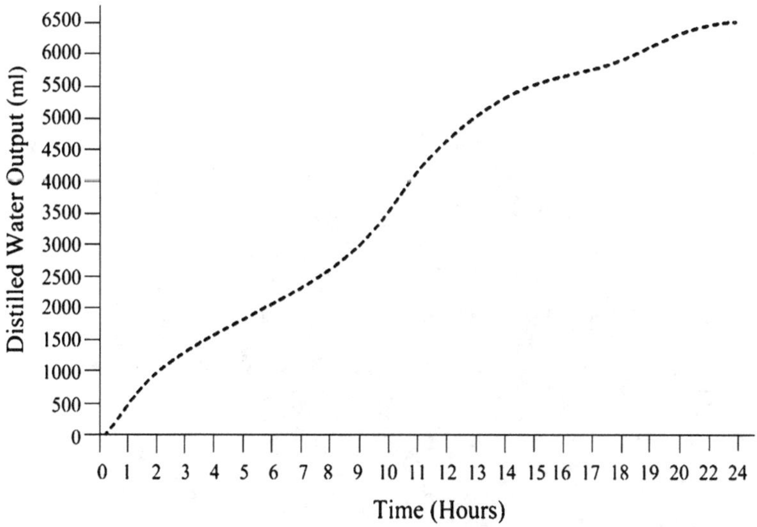

Fig. 2.4 Distilled water output vs. duration time in new solar distillation system

$$P_F = 0.7 \times C_E \times C_T \times I_S \times P_S \times C_S \quad (2.4)$$

where P_F—snow load, C_E—exposure factor, C_T—thermal factor, I_S—importance factor, P_S—50-year ground snow load, and C_S—slope factor.

The wind load and snow load, calculated from above mathematical models, can be applied in the computer-aided modeling and numerical simulation to determine if this new solar distillation system can properly function during severe weather environment. The FEA simulation and structural analysis are demonstrated in Figs. 2.5 and 2.6.

The computer-aided simulation and analysis in Figs. 2.5 and 2.6 indicate the profiles of stress and deflection in this newly designed solar distillation system. The analytic results show that the maximum stress of 13,358.53 psi on structural components in this solar system is less than the material yield strength of 36,300 psi and maximum deflection of 0.0106 in. is within material allowable deformation limit. So this solar distillation system can properly work in different severe weather environments including strong wind and heavy snow conditions.

2.3 Experiment on Solar Energy System for Water Distillation

The prototype of this new solar energy distillation system has been built and experiments have been conducted with all tested results being demonstrated as follows.

Table 2.1 displays the prototype testing result of solar radiation intensity vs. experimental duration time.

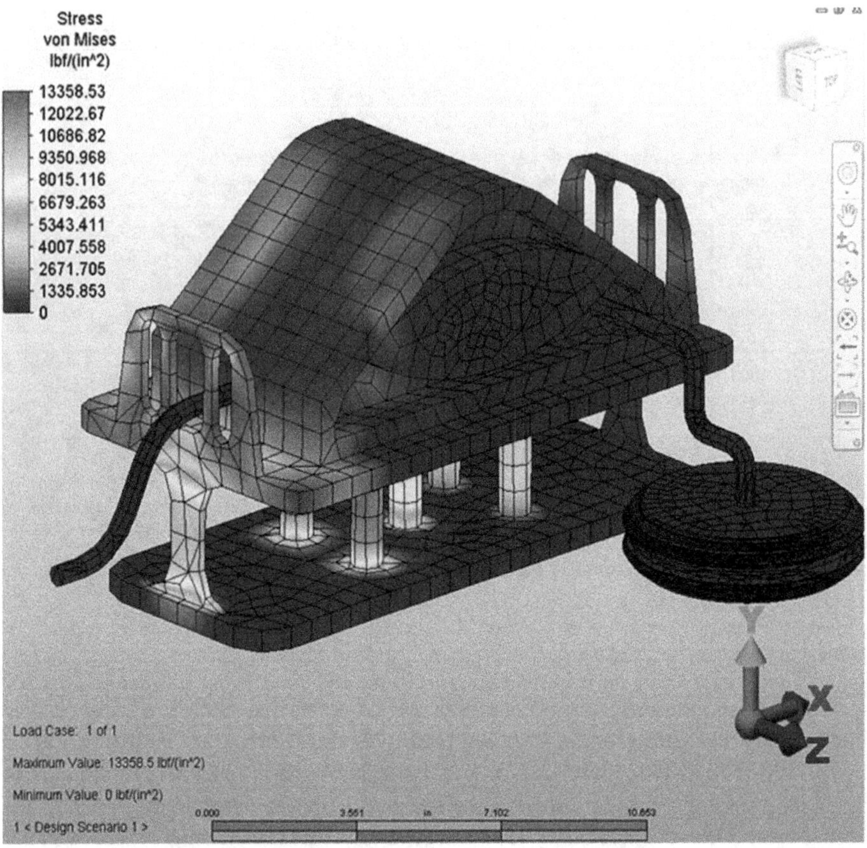

Fig. 2.5 Computer-aided simulation of stress profile in new solar distillation system

Table 2.2 presents the prototype testing result of solar distilled water output vs. experimental duration time.

The above prototype testing verifies the proper function of this system since the experimental results of solar radiation intensity vs. experimental duration time in Table 2.1 and solar distilled water output vs. experimental duration time in Table 2.2 are very close to the results stated, respectively, in Figs. 2.3 and 2.4 by computer-aided modeling and numerical simulation.

Table 2.3 shows the experiment results of maximum stress and maximum deflection in this new solar distillation energy system.

The prototype experimental results in Table 2.3 confirm the appropriate function of this system since the average maximum stress 13,358.51 psi and average maximum deflection 0.0111 in. are very similar to the results of maximum stress 13,358.53 psi and maximum deflection 0.0106 in. that are expressed, respectively, in Figs. 2.5 and 2.6 by computer-aided modeling and simulation.

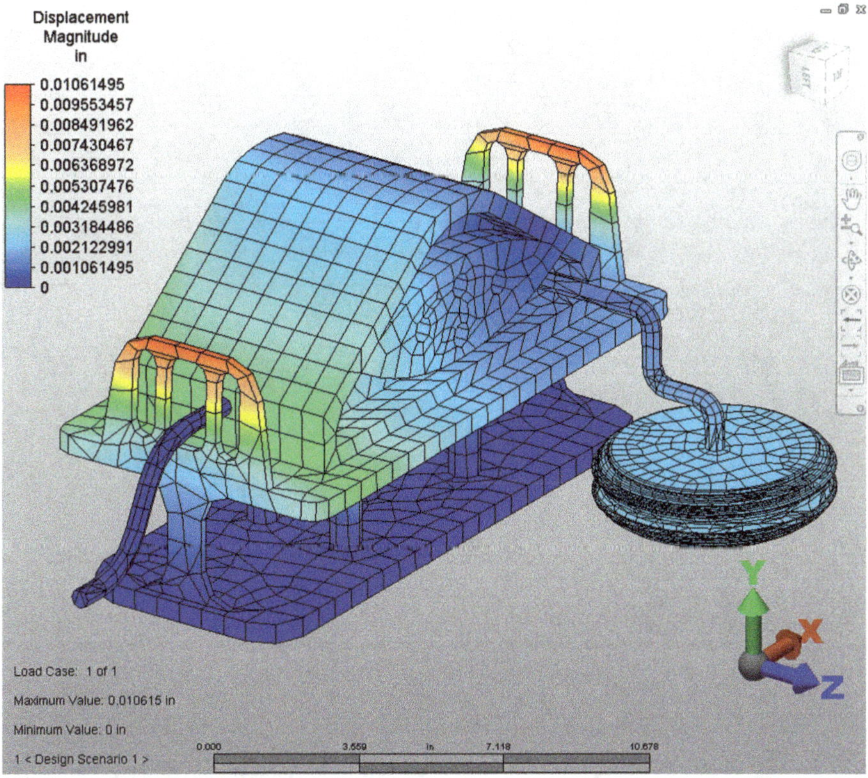

Fig. 2.6 Computer-aided simulation of deflection profile in new solar distillation system

Table 2.1 Prototype testing of solar radiation intensity vs. experimental duration time

Time (h)	Solar radiation intensity (W/m^2)
2	920
4	1,150
6	1,380
8	1,750
10	2,260
12	2,320
14	1,880
16	1,340
18	660
20	480
22	290
24	160

Table 2.2 Prototype testing of solar distilled water output vs. experimental duration time

Time (h)	Distilled water output (ml)
2	950
4	1,480
6	1,940
8	2,480
10	3,550
12	4,780
14	5,450
16	5,680
18	5,880
20	6,210
22	6,370
24	6,460

Table 2.3 Prototype testing of maximum stress and maximum deflection in this new solar distillation energy system

Number of experiment	Maximum stress (psi)	Maximum deflection (in.)
1	13,358.48	0.0111
2	13,358.55	0.0104
3	13,358.52	0.0115
4	13,358.49	0.0112
5	13,358.55	0.0108
6	13,358.46	0.0114
7	13,358.54	0.0115
8	13,358.51	0.0105
9	13,358.48	0.0111
10	13,358.44	0.0115
11	13,358.56	0.0108
12	13,358.51	0.0104
13	13,358.47	0.0114
14	13,358.44	0.0112
15	13,358.56	0.0115
16	13,358.61	0.0103
17	13,358.44	0.0114
18	13,358.48	0.0112
19	13,358.52	0.0115
20	13,358.49	0.0104
Average	13,358.51	0.0111

2.4 Discussion and Future Improvement of Solar Energy System for Water Distillation

The filtration process can take away many environmental pollutants caused by manufacturing process and improper disposal of industrial wastes that have been detected in the drinking water supply. Solar water distillation is one of the most effective technologies for removing all such pollutants. There are a few disfavors in solar water distillation systems including the capital expense of solar unit in the front. To reduce the front cost of this new solar water distillation unit, some modifications are planned, i.e., simplify the geometrical design layout, modify the heat exchanger cell for better thermal efficiency, and analyze to select appropriate materials to bring down the unit cost to further improve this new solar water distillation system.

Wind Power Turbine System

3

3.1 Design of New Wind Power Turbine System

The wind power turbines, with horizontal or vertical axis types, are technical equipment that convert wind kinetic energy to mechanical energy used to generate electricity. The applications range from residential appliance battery charge using small wind power turbines to commercial electricity production using large wind power turbine system. Wind power turbines are developed not only to utilize the green wind energy for different energy enterprises but also to protect environment from air pollution. Computer-aided aerodynamic modeling and analysis can be applied to define the height of wind power turbine system, determine turbine blade geometry, specify turbine control system, and choose the number of turbine blades. The major turbine components include rotor with blades installed to change wind kinetic energy to slowly rotated mechanical energy, electrical generator with drive gearbox and electronics controller to change from slow-inlet shaft rotation to high-outlet shaft rotation for electricity generation, and all structural components to support wind power turbine system.

Two new wind power turbine systems, based on author's new research, were designed through computer-aided modeling, numerical simulation, and prototype testing. Figures 3.1, 3.2, 3.3, and 3.4 display the 3D models of these two new wind power turbines and some turbine components including driver gear and blade.

In design and development of wind power turbine system, the segmental element theory is applied in computer-aided design and engineering analysis to determine 3D complex geometry of turbine components, analyze the dynamic forces on a full turbine blade by dividing a blade into multiple small segments to find the load profile in different blade elements, and verify the turbine performance. The full load exerted to the rotor can be determined by integrating the segmental loads for all turbine blade elements. The maximum rising force can be found when turbine blade moves in opposite direction to the wind direction. The critical parameters in turbine

J. Zheng Li, *CAD, 3D Modeling, Engineering Analysis, and Prototype Experimentation*, DOI 10.1007/978-3-319-05921-1_3

Fig. 3.1 Wind power turbine system (new design 1)

Fig. 3.2 Wind power turbine system (new design 2)

Fig. 3.3 Driver gear design in new wind power turbine system

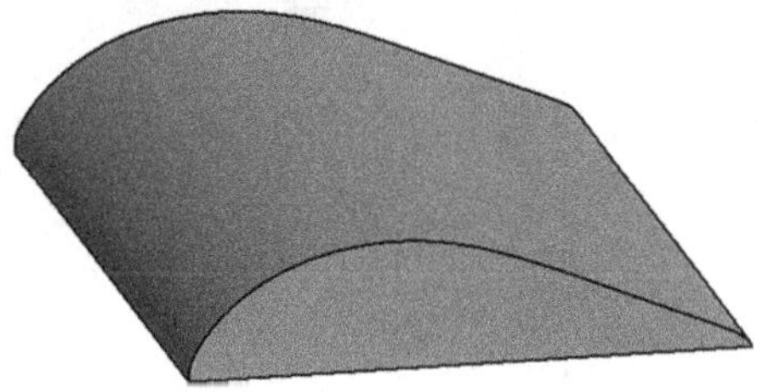

Fig. 3.4 Turbine blade design in new wind turbine system

design and development are oriented angle β between arriving air and blade orientation, forces acted at the quarter of chord from blade head edge, pitching torque caused by dynamic forces of coming air, tangential force of coming air to draw the blade, and normal force of coming air to rise the blade. The computer-aided simulation and prototype testing of these two new wind power turbine systems are described in the following chapters.

3.2 Computer-Aided Simulation of Wind Power Turbine System

The computer-aided simulation and analysis of lifting and dragging forces in wind power turbine energy system can be applied to determine these targeted output forces by analyzing the aerodynamic performance around the wind turbine blade. All the analytic results can be used to guide wind turbine system design, improve turbine functionality, and optimize turbine system for best performance. Equation (3.1) can be applied to specify the lifting force on each turbine blade (Kundu and Cohen 2008):

$$N_{\mathrm{LIFT}} = \frac{\rho_{\mathrm{AIR}} \times A_{\mathrm{NET}} \times D_{\mathrm{LIFT}} \times S_{\mathrm{RESULTANT}}^2}{2} \tag{3.1}$$

Here, ρ_{AIR}—density of incoming air, A_{NET}—net surface area of turbine blade, D_{LIFT}—lifting coefficient, and $S_{\mathrm{RESULTANT}}$—resultant speed.

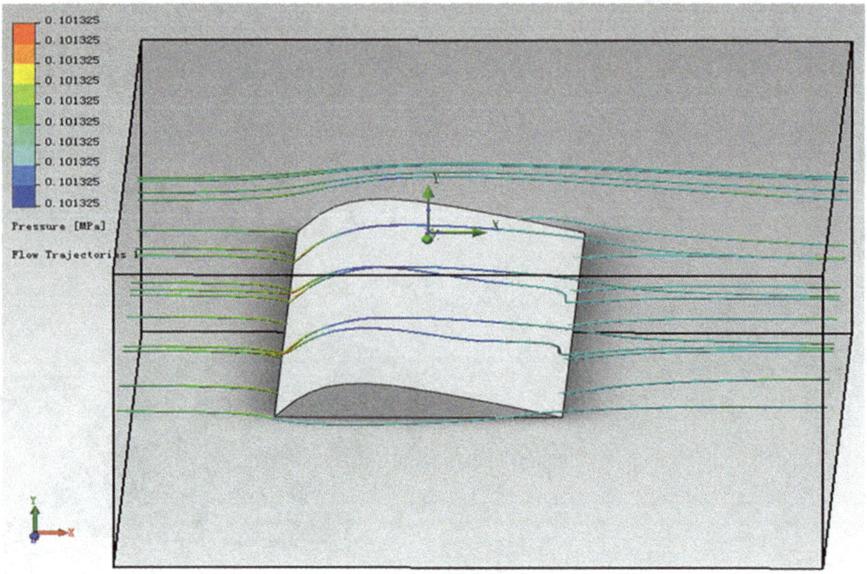

Fig. 3.5 Computer-aided simulation of airflow over wind power turbine blade

The full wind power can be determined by the following equation (Kundu and Cohen 2008):

$$P_{\mathrm{FULL}} = \frac{D_{\mathrm{POWER}} \times \rho_{\mathrm{AIR}} \times A_{\mathrm{SWEPT}} \times S_{\mathrm{WIND}}^{3}}{2} \tag{3.2}$$

Here, D_{POWER}—power coefficient, A_{SWEPT}—swept area in wind power turbine, and S_{WIND}—wind speed.

The turbine blade material is made of fiber glass composite with high strength and low density. The complex geometrical contour design in wind power system is assisted by computer-aided 3D modeling and aerodynamic analysis in determining turbine lifting force and turbine full power has been performed through computational simulation and finite element analysis. Figure 3.5 demonstrates the computer-aided simulation of airflow over wind power turbine blade.

Computer-aided design and analysis is the computational methodology to model the system design and simulate the product function, system structure, and produced stress and deflection profiles in designed system. All the simply defined individual elements can be calculated through computational simulation and analytic results for full system can be finally determined. Figures 3.6, 3.7, 3.8, 3.9, 3.10, and 3.11 represent the stress and deflection profiles for some critical components in this new wind power turbine system through computer-aided modeling and simulation.

The computer-aided simulation and analysis in Figs. 3.6 and 3.7 exhibit the stress and deflection profiles of turbine blade in this newly designed system.

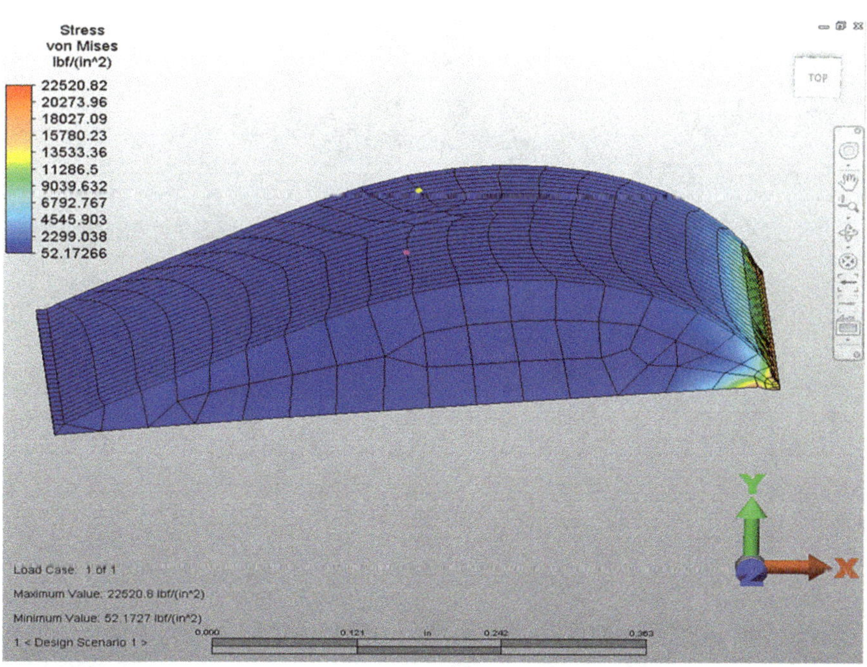

Fig. 3.6 Computational modeling of stress profile in turbine blade

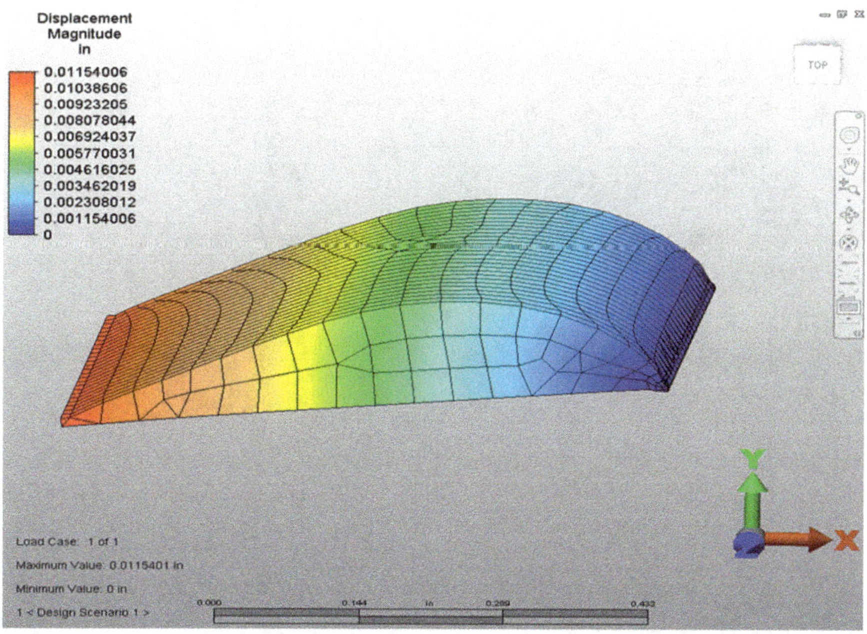

Fig. 3.7 Computational modeling of deflection profile in turbine blade

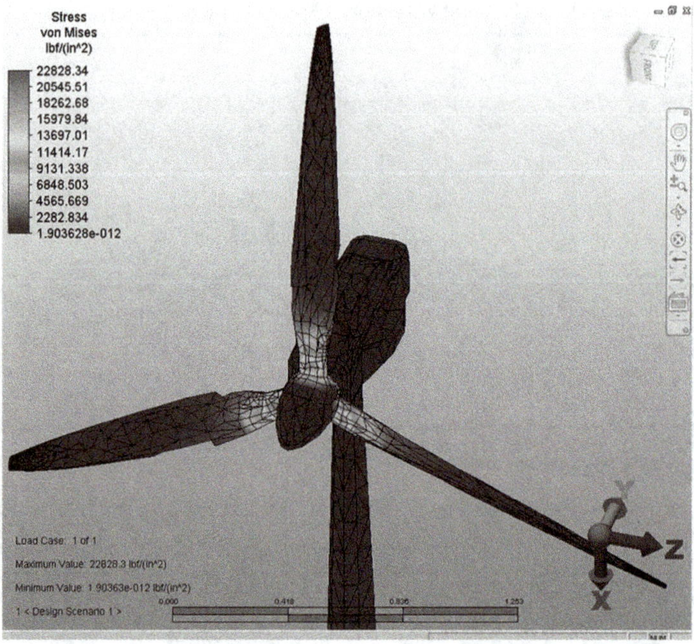

Fig. 3.8 Computational modeling of stress profile in wind power turbine system 1

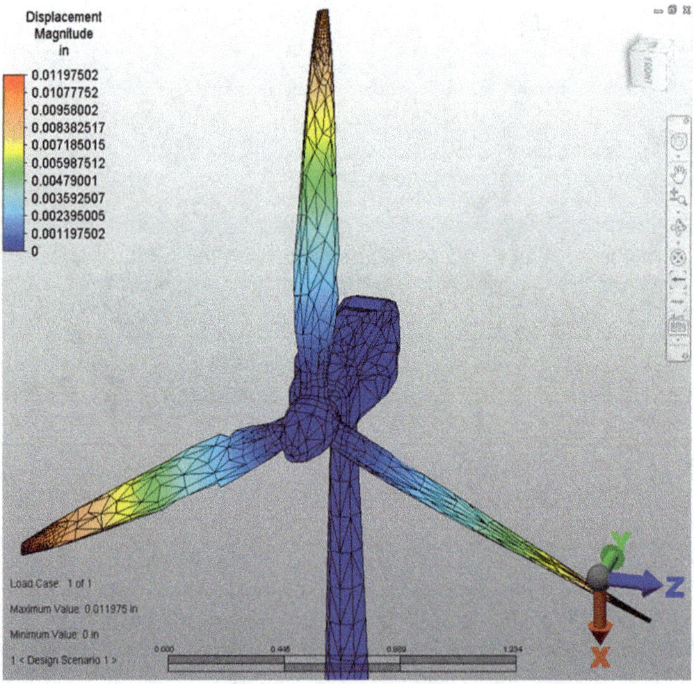

Fig. 3.9 Computational modeling of deflection profile in wind power turbine system 1

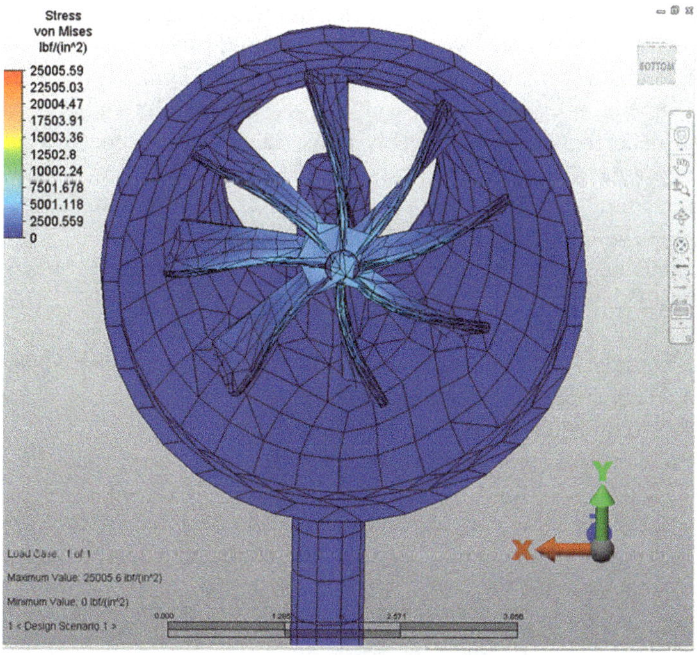

Fig. 3.10 Computational modeling of stress profile in wind power turbine system 2

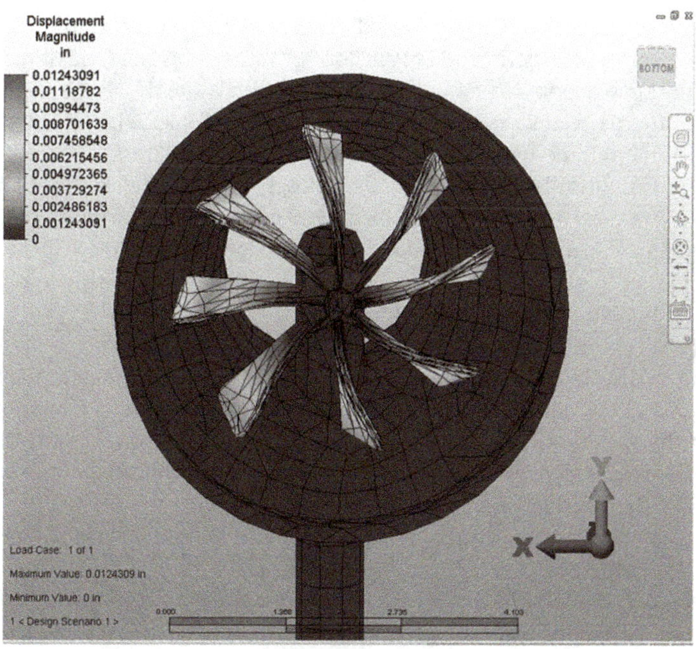

Fig. 3.11 Computational modeling of deflection profile in wind power turbine system 2

The analytic results show that the maximum stress of 22,520.82 psi in turbine blade is less than the material yield strength of 36,300 psi and maximum deflection of 0.01154 in. is within material allowable deformation limit.

The computer-aided simulation and analysis in Figs. 3.8 and 3.9 exhibit the stress and deflection profiles of newly designed turbine system_1. The analytic results present that the maximum stress of 22,828.34 psi in turbine system_1 is less than the material yield strength of 36,300 psi and maximum deflection of 0.01198 in. is within material allowable deformation limit.

The computer-aided simulation and analysis in Figs. 3.10 and 3.11 demonstrate the stress and deflection profiles of newly designed turbine system_2. The analytic results present that the maximum stress of 25,005.59 psi in turbine system_2 is less than the material yield strength of 36,300 psi and maximum deflection of 0.01243 in. is within material allowable deformation limit.

The above computer-aided 3D modeling and numerical simulation display that the maximum stresses on these important parts are all lower than the material strength and all related maximum component deformations meet material deflection spec. Based on above analytic results, this newly developed wind power energy system shows its proper systematic design, feasible function, and reliable structure in the green and sustainable energy applications.

3.3 Experiment on Wind Power Turbine System

New wind power energy system has been prototyped and tested to examine and assert the computer-aided analysis and numerical simulation results. Table 3.1 shows the prototype experimental results of maximum stress and maximum deflection on turbine blade in this new wind power turbine system.

The prototype testing results of turbine blade in Table 3.1 prove the normal function of the turbine blade since the average maximum stress 22,520.85 psi and average maximum deflection 0.0112 in. are closely equal to the results of maximum stress 22,520.82 psi and maximum deflection 0.0115 in. that are exhibited, respectively, in Figs. 3.6 and 3.7 by computer-aided modeling and numerical simulation.

Table 3.2 presents the testing results of maximum stress and maximum deflection of wind power turbine energy system 1.

The prototype testing results of new wind power turbine energy system 1 in Table 3.2 verify the proper function of system 1 because the average maximum stress 22,828.29 psi and average maximum deflection 0.0116 in. are approximately equal to the results of maximum stress 22,828.34 psi and maximum deflection 0.0120 in. that are depicted, respectively, in Figs. 3.8 and 3.9 by computer-aided modeling and numerical simulation.

Table 3.3 expresses the experimental results of maximum stress and maximum deflection of wind power turbine energy system 2.

The prototype experimental results of new wind power turbine energy system 2 in Table 3.3 confirm the appropriate function of system 2 since the average

Table 3.1 Prototype experiment of maximum stress and maximum deflection on turbine blade in this new wind power energy system

Number of experiment	Maximum stress (psi)	Maximum deflection (in.)
1	22,520.88	0.0118
2	22,520.84	0.0114
3	22,520.78	0.016
4	22,520.80	0.0109
5	22,520.89	0.0108
6	22,520.89	0.0106
7	22,520.88	0.0118
8	22,520.86	0.0111
9	22,520.79	0.0105
10	22,520.78	0.0108
11	22,520.87	0.0108
12	22,520.89	0.0108
13	22,520.92	0.0112
14	22,520.81	0.0106
15	22,520.80	0.0108
16	22,520.86	0.0111
17	22,520.93	0.0108
18	22,520.88	0.0107
19	22,520.78	0.0108
20	22,520.83	0.0116
Average	22,520.85	0.0112

Table 3.2 Prototype testing results of maximum stress and maximum deflection of wind power turbine energy system 1

Number of experiment	Maximum stress (psi)	Maximum deflection (in.)
1	22,828.26	0.0118
2	22,828.35	0.0122
3	22,828.25	0.0112
4	22,828.38	0.0124
5	22,828.26	0.0112
6	22,828.24	0.0111
7	22,828.31	0.0113
8	22,828.38	0.0123
9	22,828.25	0.0111
10	22,828.24	0.0112
11	22,828.27	0.0117
12	22,828.36	0.0123
13	22,828.29	0.0114
14	22,828.25	0.0112
15	22,828.35	0.0115
16	22,828.25	0.0122
17	22,828.26	0.0112
18	22,828.37	0.0112
19	22,828.24	0.0121
20	22,828.31	0.0122
Average	22,828.29	0.0116

Table 3.3 Prototype testing results of maximum stress and maximum deflection of wind power turbine energy system 2

Number of experiment	Maximum stress (psi)	Maximum deflection (in.)
1	25,005.65	0.0129
2	25,005.61	0.0122
3	25,005.55	0.0132
4	25,005.57	0.0129
5	25,005.68	0.0121
6	25,005.67	0.0133
7	25,005.54	0.0130
8	25,005.66	0.0133
9	25,005.68	0.0121
10	25,005.57	0.0132
11	25,005.66	0.0122
12	25,005.68	0.0131
13	25,005.65	0.0133
14	25,005.54	0.0130
15	25,005.57	0.0122
16	25,005.68	0.0121
17	25,005.67	0.0132
18	25,005.65	0.0133
19	25,005.57	0.0123
20	25,005.55	0.0122
Average	25,005.62	0.0128

maximum stress 25,005.62 psi and average maximum deflection 0.0128 in. are almost equal to the results of maximum stress 25,005.59 psi and maximum deflection 0.0124 in. that are depicted, respectively, in Figs. 3.10 and 3.11 by computer-aided modeling and numerical simulation.

3.4 Discussion and Future Improvement on Wind Power Turbine System

In the near future, wind energy will become one of the most cost-effective and environment-friendly sources of electrical power. In wind power energy system, a turbine is connected to a mechanical electrical generator so that the electricity can be generated by natural wind energy. The piezoelectric materials can be mechanically formed into different geometrical blades and connected to the piezoelectric actuators to make electrical power by wind power energy (wind pressure). Because the wind power energy system is a passive system (not a mechanically driven system), it will be less likely broken down and, therefore, no frequent and costly repair is usually required. In order to further improve this new wind power energy system, the following modifications will be planned: (1) modify the turbine blade structure to keep wind flow passing turbine blade more consistent, (2) simplify the turbine structure to have less materials required to build wind turbine to lower the cost, (3) rectify this new turbine system so that it will cost less energy for unit operation, (4) amend this new design for easier system installation, (5) and improve the airflow pathway near diameter of the turbine blades to maintain a low-pressure region that increases wind flow to the turbine for high system efficiency.

Solar Panel Tracking System 4

4.1 Design of Solar Panel Tracking System

A newly developed solar tracking system is introduced with its adjustable orientation angle to track daily sunlight in 3D directions in order to receive the maximum solar energy via systematic photovoltaic arrays. The driving mechanisms designed with gear train unit in this solar tracking system can orientate solar system between east and west and a rotating table can orientate solar unit between north and south. The prototype of this new solar tracking system is depicted in Figs. 4.1 and 4.2.

Figure 4.1 shows the prototype of this new solar tracking system that has an orientation rack and orientation base (with stepper motor and gear reducer inside) installed to rotate solar panel frame in the eastern-western direction and has motor driving system, base plate, and orientation plate mounted to keep solar panel frame gradually rotating to follow sun's yearly travel in north-south orbit. Several detecting sensors are installed to trace the sunlight through receiving varied current signals from photodiodes in order to automatically manipulate the solar panel rotation in eastern-western and northern-southern 3D directions.

4.2 Computer-Aided Simulation of Solar Panel Tracking System

This new solar tracking system is designed based on analysis of computer-aided modeling and numerical simulation. The 3D modeling of the solar tracking system is performed by 3D CAD software and structure analysis is carried out by FEA technique. The structure analysis includes validating the functionality and strength of driving system in east-west and north-south solar panel orientation.

J. Zheng Li, *CAD, 3D Modeling, Engineering Analysis, and Prototype Experimentation*, DOI 10.1007/978-3-319-05921-1_4

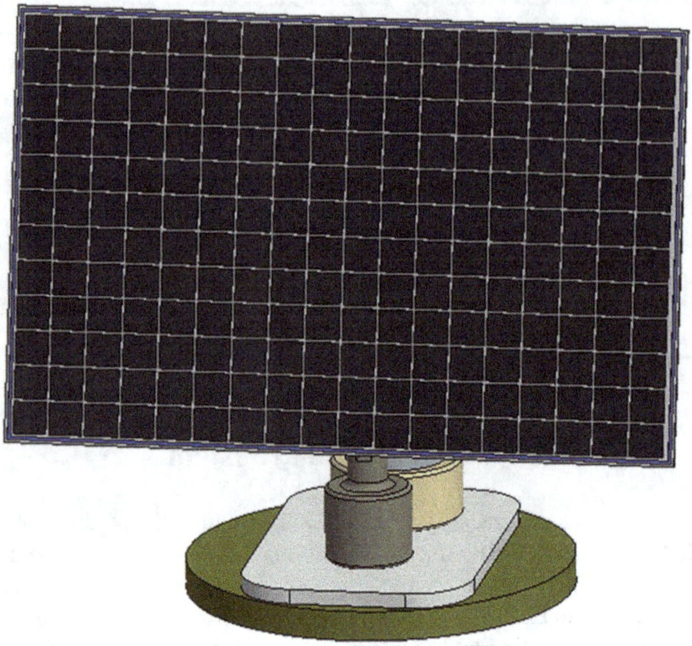

Fig. 4.1 Solar panel system front view

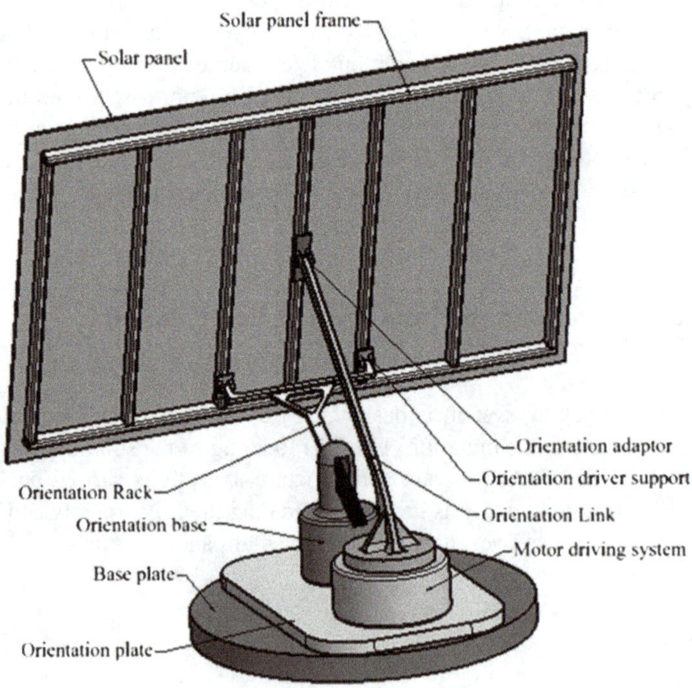

Fig. 4.2 Solar panel system rear view

1. Calculation of wind force:

The wind load required should be considered in the solar tracking system design to make sure that all the functioning parts, such as panel frame, lateral channel beam, longitudinal channel beam, orientation adaptor, orientation rack, orientation base, orientation driver support, and orientation link, will still work under the maximum wind load in worst conditions. The wind load equation can be expressed as follows (Mehta and Coulburne 2010):

$$F_{\text{wind}} = A_{\text{project}} \times P_{\text{wind}} \times D_{\text{drag}} \tag{4.1}$$

The wind pressure P_{wind} can be defined by the following equation:

$$P_{\text{wind}} = 0.00256 \times V_{\text{wind}}^2 \tag{4.2}$$

Here, F_{wind}—load caused by wind (lbf), A_{project}—projected surface area of solar panel at different orientation (ft^2), P_{wind}—pressure caused by wind (psf), D_{drag}—coefficient of drag, and V_{wind}—speed of wind (mph).

2. Calculation of gear-train force in solar panel orientation:

To keep the gears from damage during solar panel tracking system operation, the gear train must be able to handle the resultant force from force of wind, weight of solar panel and hardware, and other related frictional forces. The resultant force can be determined as follows (Mehta and Coulburne 2010):

$$\begin{aligned} T_{\text{panel}}(\text{torque to orientate the solar panel}) &= N_{\text{d}} \times 0.5 \times d_{\text{p}} \\ &= F_{\text{R}} \times 0.5 \times D_{\text{gear}} \end{aligned} \tag{4.3}$$

$$F_{\text{R}} = W_{\text{total weight}} \times C_{\text{f}}$$

Here, N_{d}—force to drive gear (N), d—gear pitch diameter, F_{R}—orientation force (N), D_{O}—outside gear diameter, W_{TW}—total weight of solar panel system (kg), and C_{f}—friction coefficient of different contact materials. The force to drive gears changes when solar panel orientates to the different angles which can be determined through computer-aided modeling and simulation.

Figures 4.3, 4.4, 4.5, 4.6, 4.7, 4.8, 4.9, 4.10, 4.11, 4.12, 4.13, 4.14, 4.15, 4.16, 4.17, 4.18, 4.19, 4.20, 4.21, 4.22, 4.23, 4.24, 4.25, 4.26, 4.27, 4.28, and 4.29 show the 3D part models, stress profile, and deflection profile of critical components in this new solar tracking system.

The computer-aided simulation and analysis in Figs. 4.4 and 4.5 demonstrate the stress and deflection profiles of newly designed solar panel system. The analytic results present that the maximum stress of 2,267.076 psi in this solar panel system is less than the material yield strength of 36,300 psi and maximum deflection of 0.00507 in. is within material allowable deformation limit.

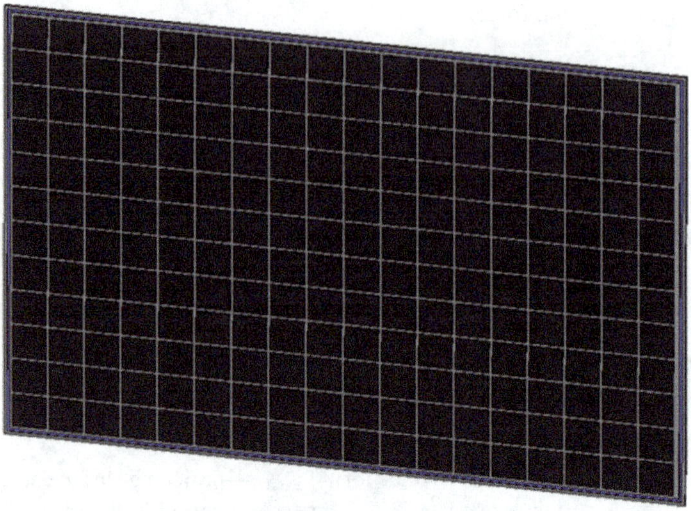

Fig. 4.3 Solar panel

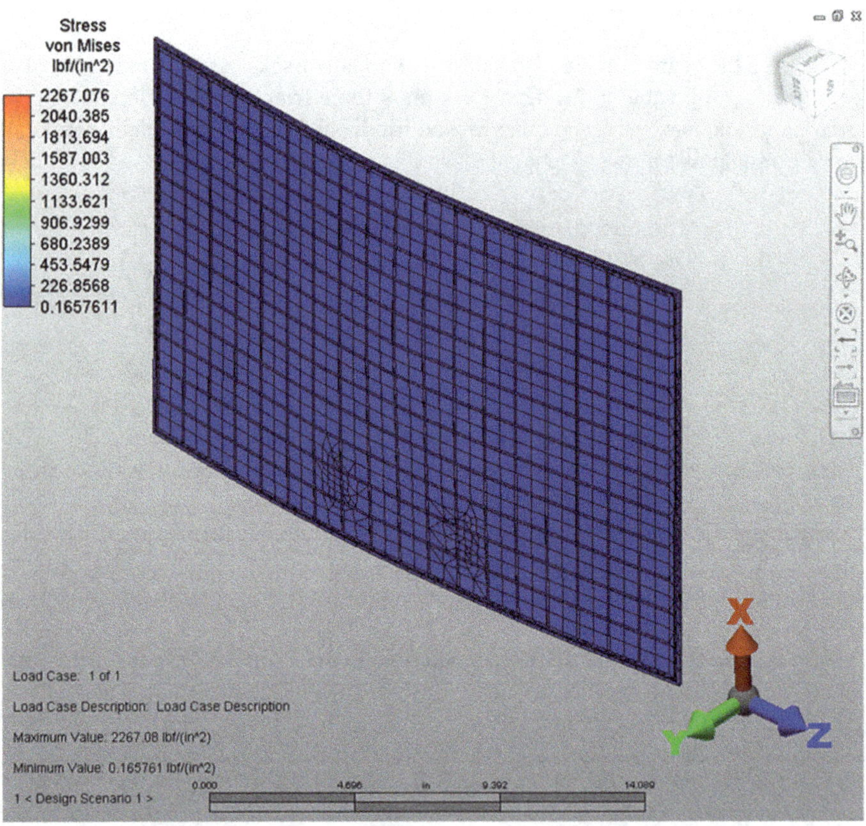

Fig. 4.4 Stress profile in solar panel

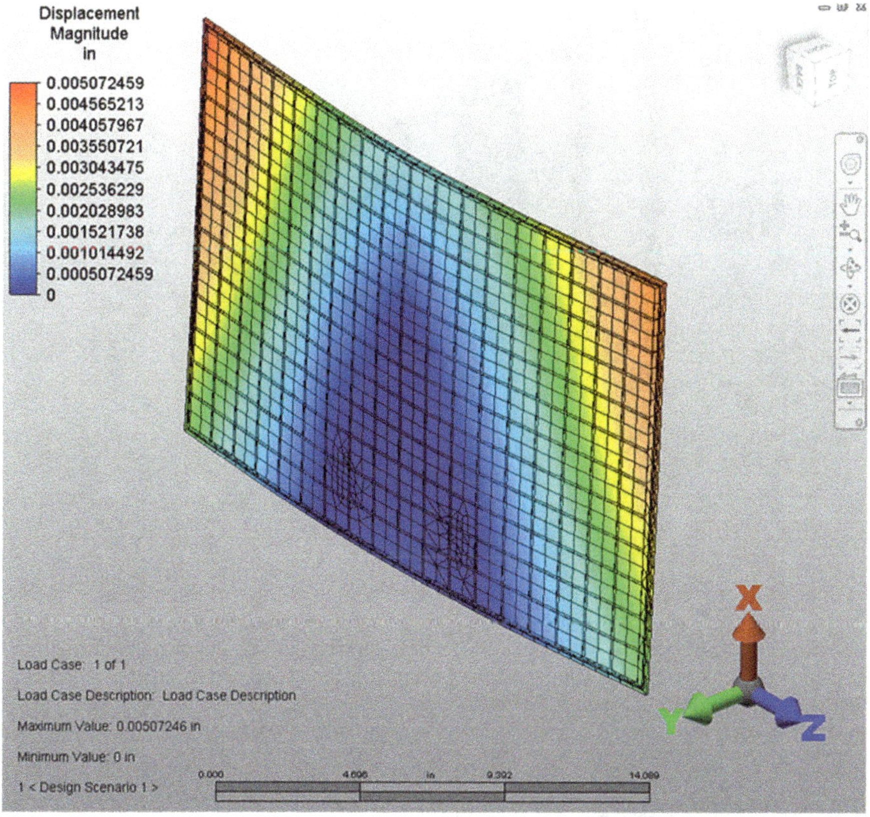

Fig. 4.5 Deflection profile in solar panel

Fig. 4.6 Solar panel frame

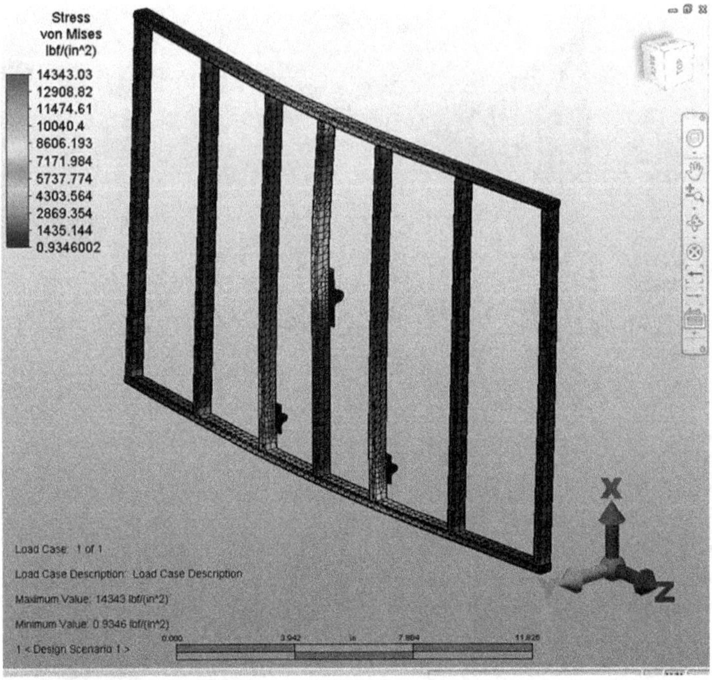

Fig. 4.7 Stress profile in panel frame

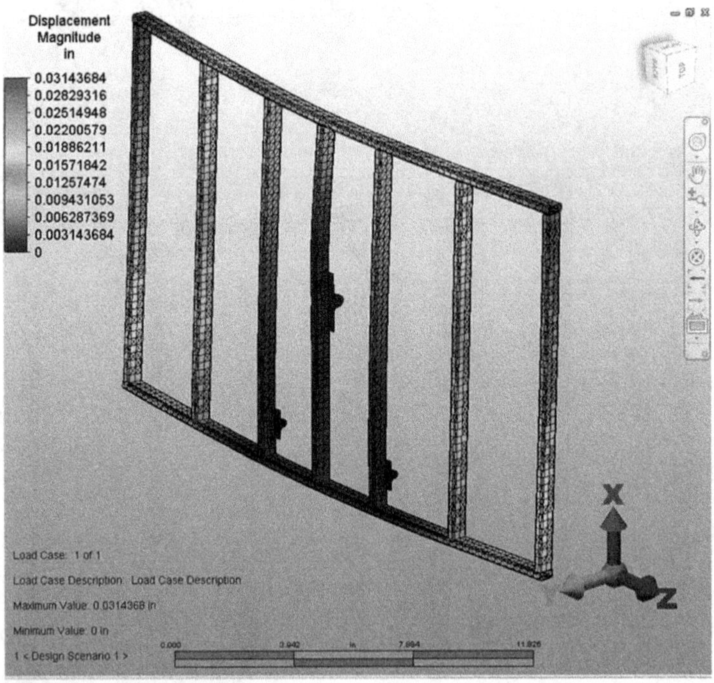

Fig. 4.8 Deflection profile in panel frame

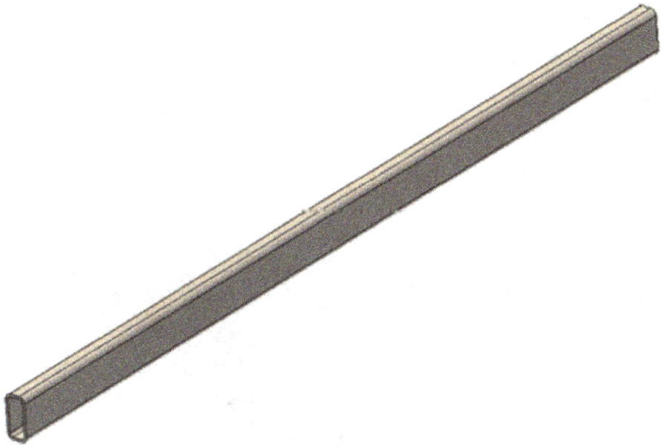

Fig. 4.9 Lateral channel beam

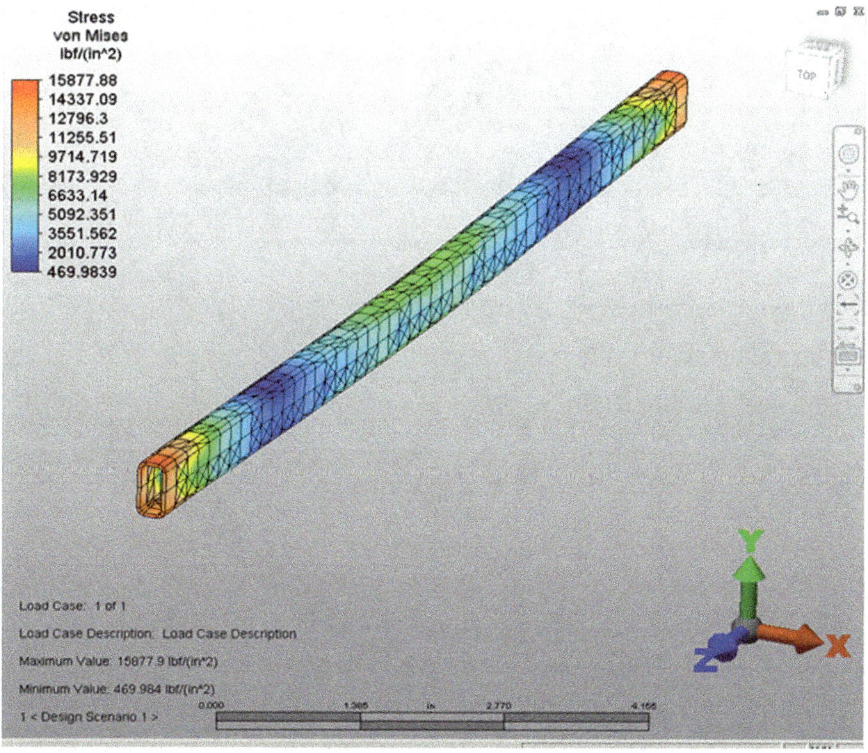

Fig. 4.10 Stress profile in lateral channel beam

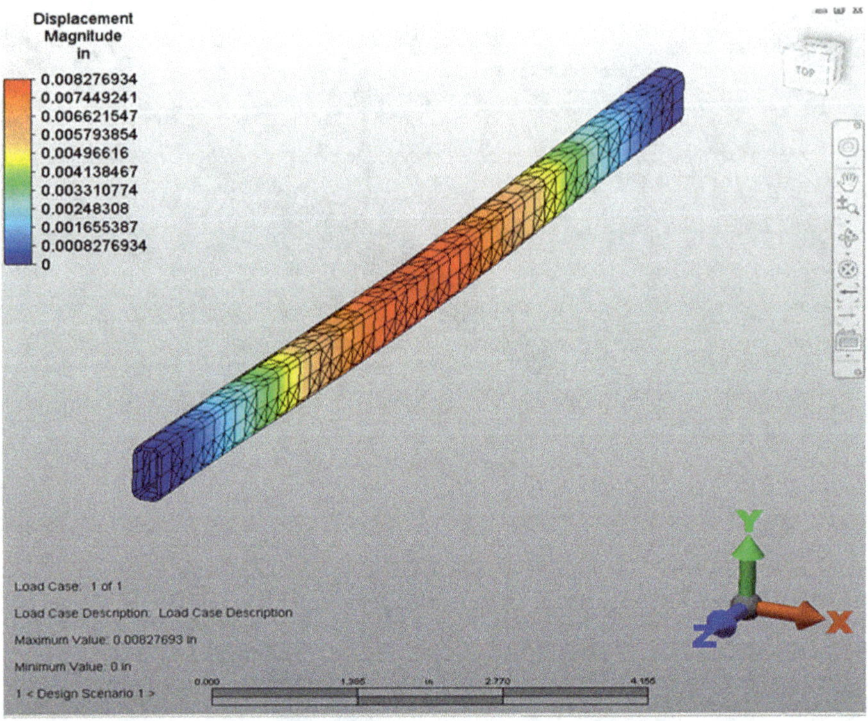

Fig. 4.11 Deflection profile in lateral channel beam

Fig. 4.12 Longitudinal channel beam

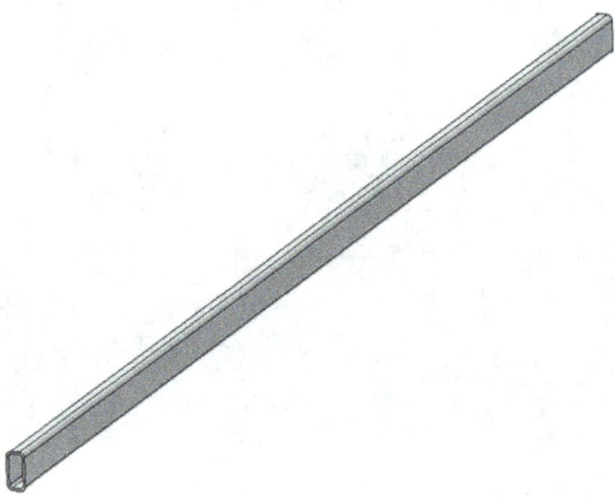

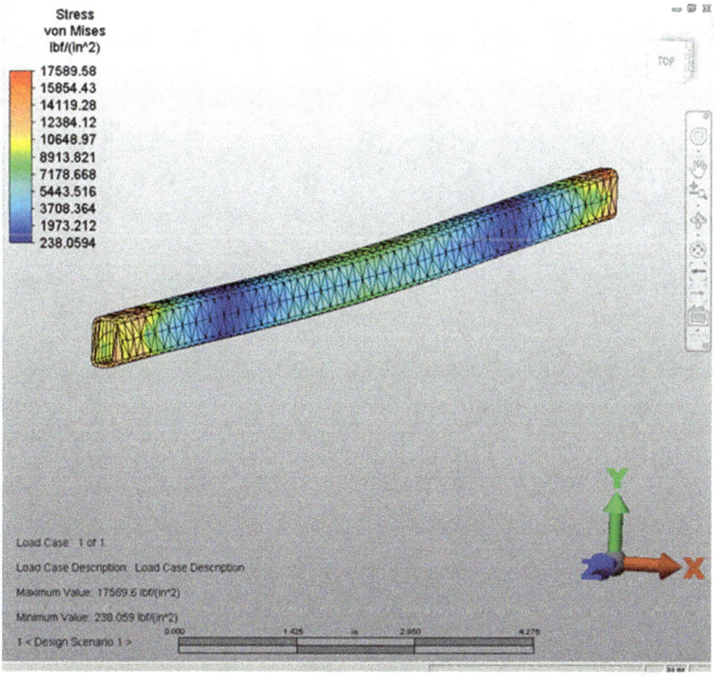

Fig. 4.13 Stress profile in longitudinal channel beam

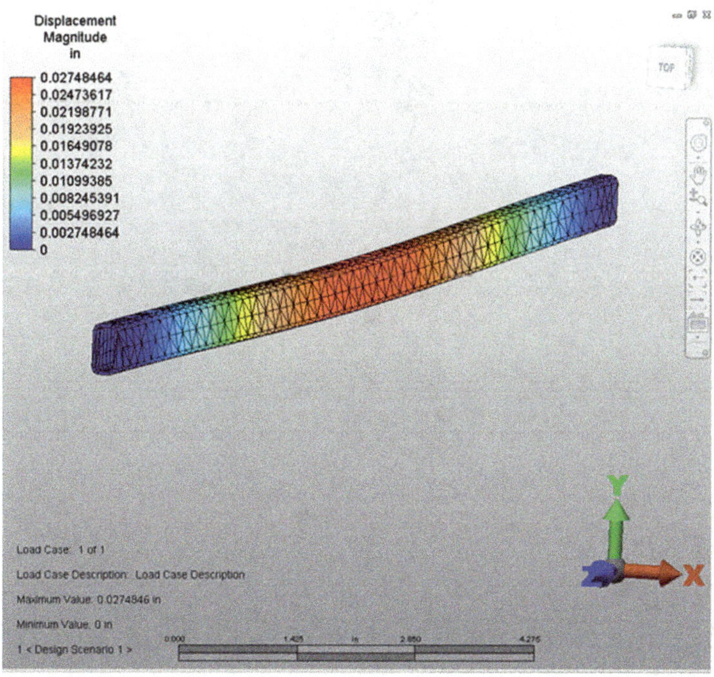

Fig. 4.14 Deflection profile in longitudinal channel beam

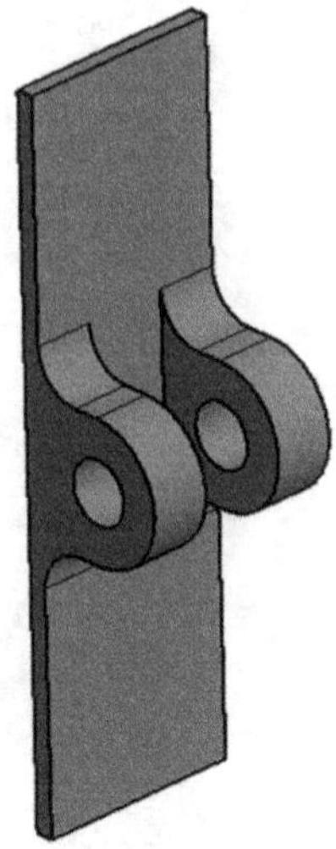

Fig. 4.15 Orientation adaptor

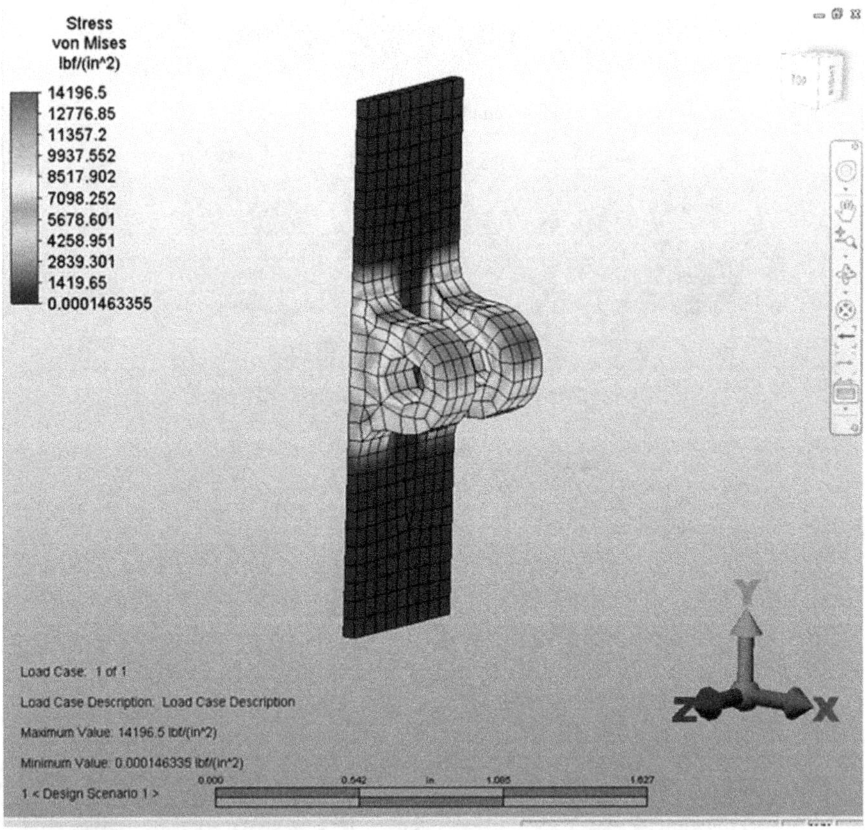

Fig. 4.16 Stress profile in orientation adaptor

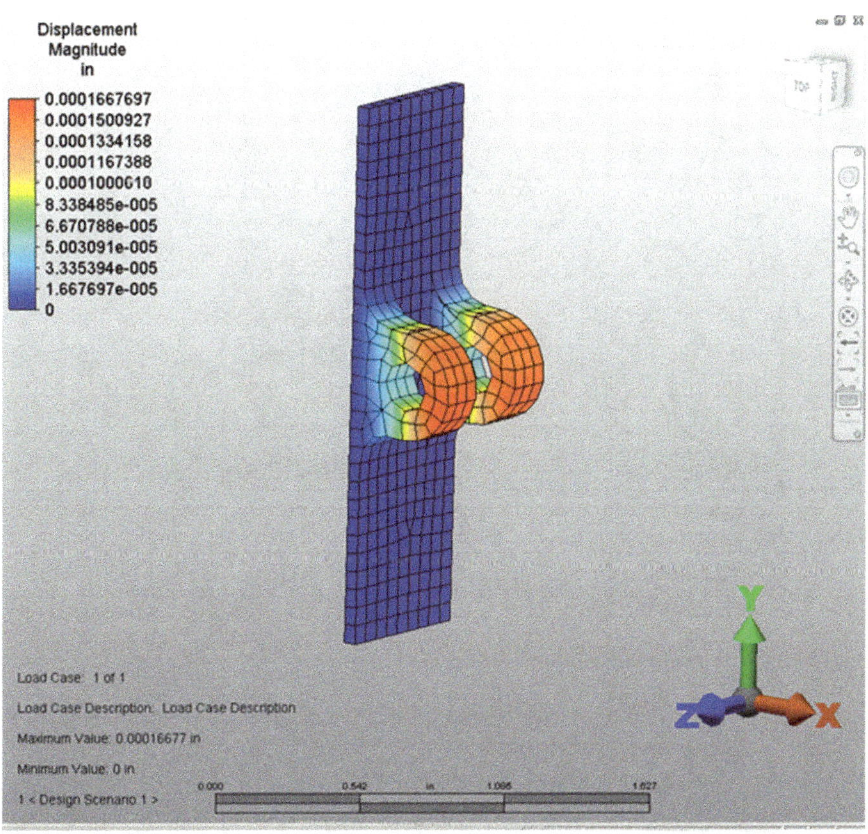

Fig. 4.17 Deflection profile in orientation adaptor

Fig. 4.18 Orientation rack

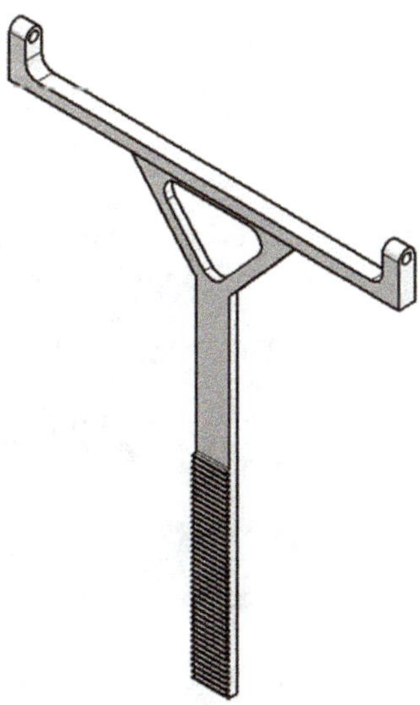

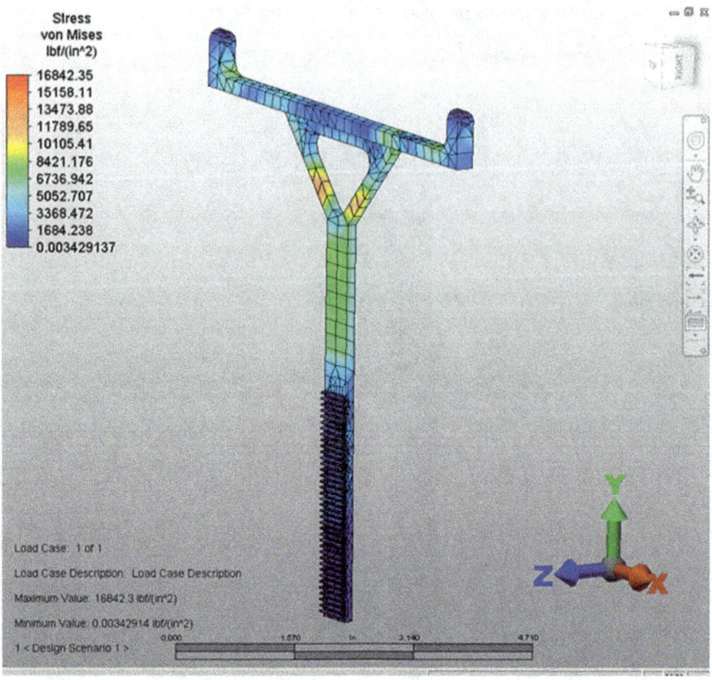

Fig. 4.19 Stress profile in orientation rack

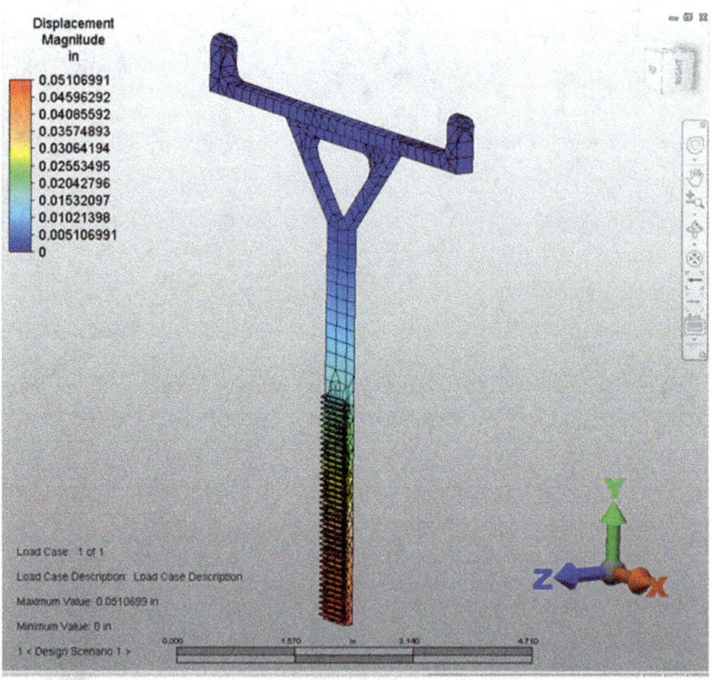

Fig. 4.20 Deflection profile in orientation rack

Fig. 4.21 Orientation base

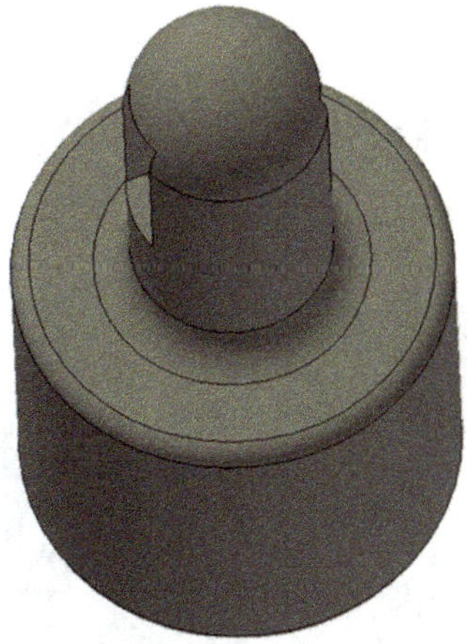

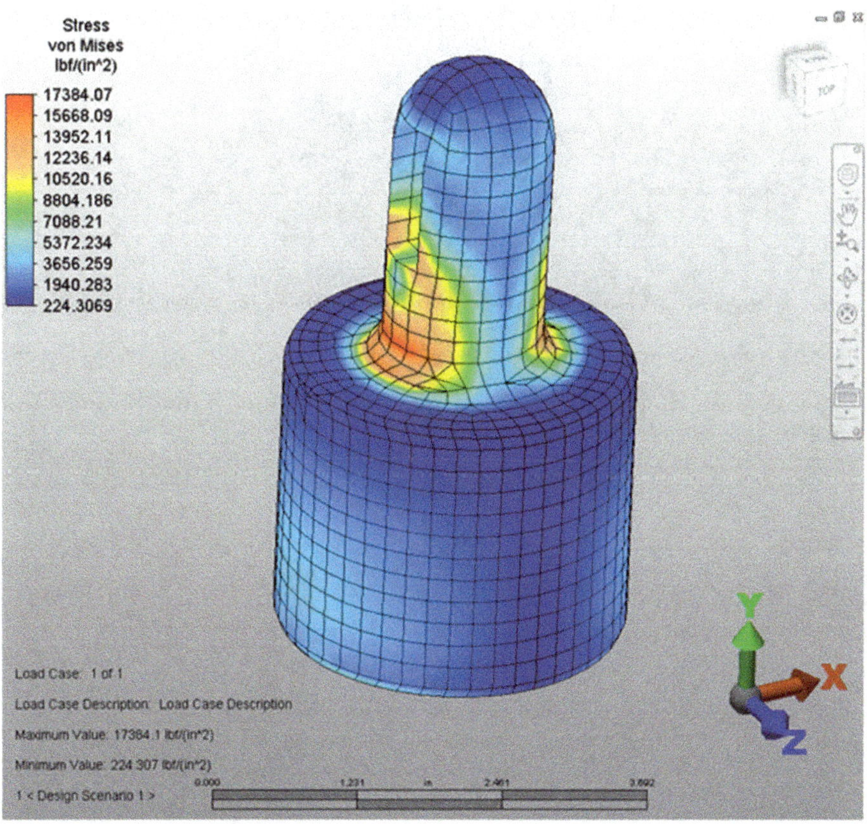

Fig. 4.22 Stress profile in orientation base

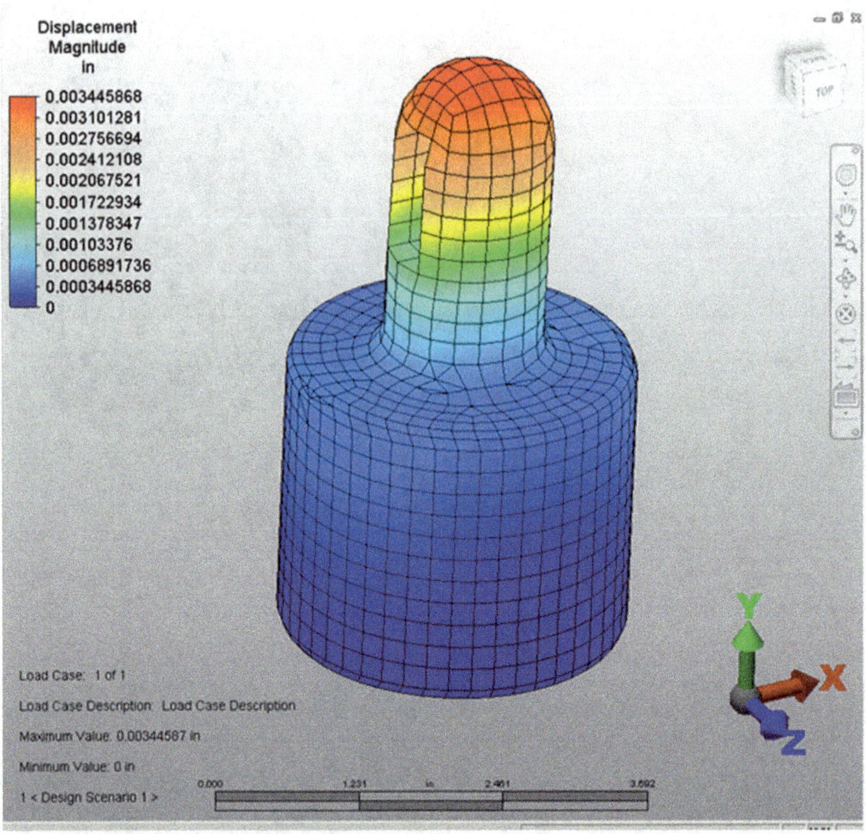

Fig. 4.23 Deflection profile in orientation base

Fig. 4.24 Orientation driver support

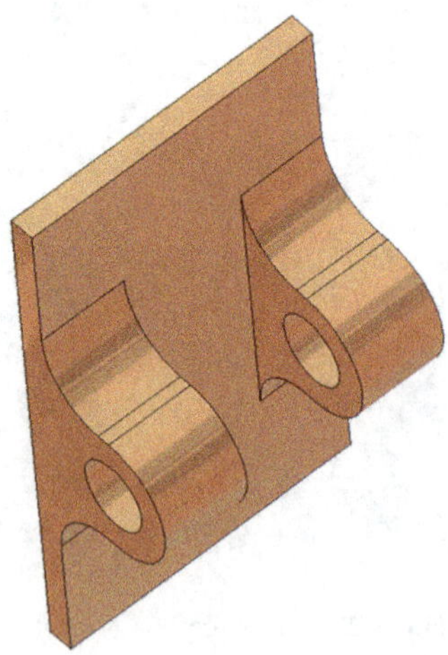

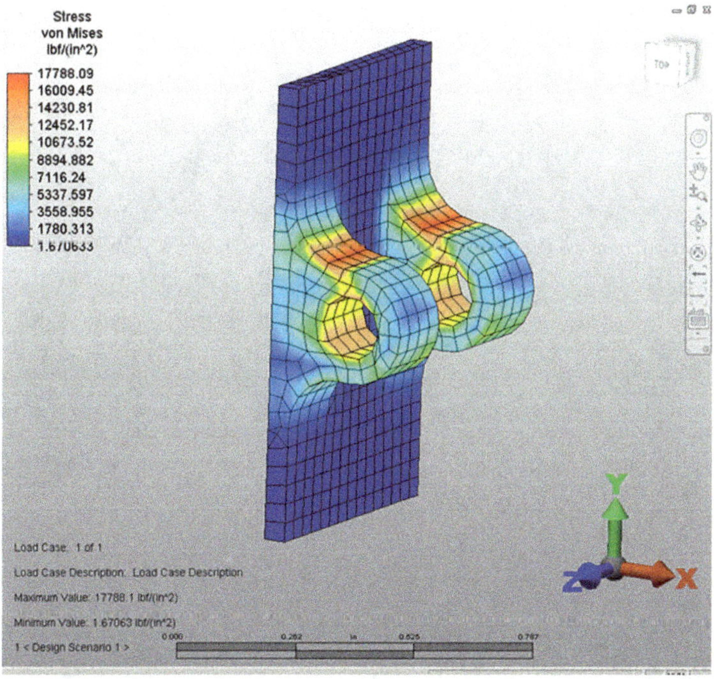

Fig. 4.25 Stress profile in orientation driver support

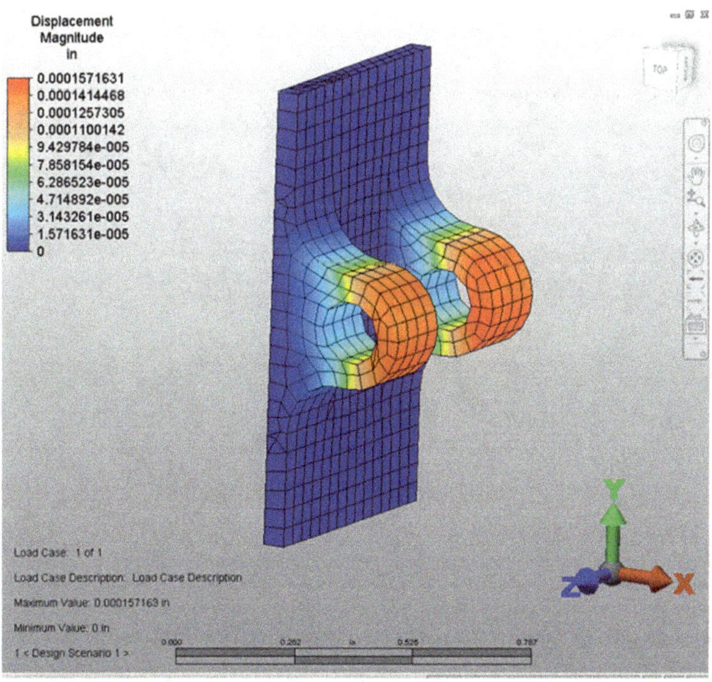

Fig. 4.26 Deflection profile in orientation driver support

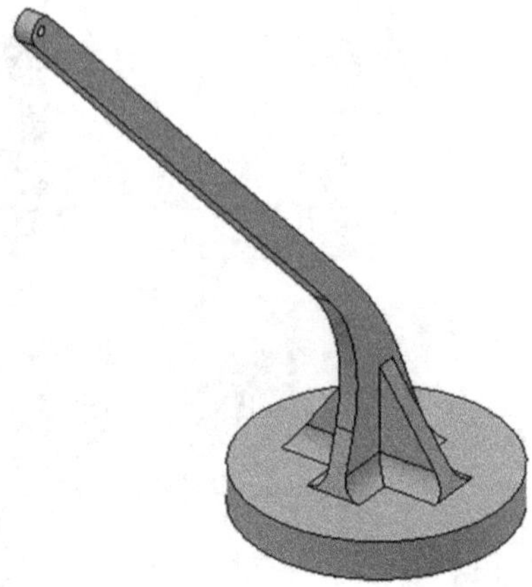

Fig. 4.27 Orientation link

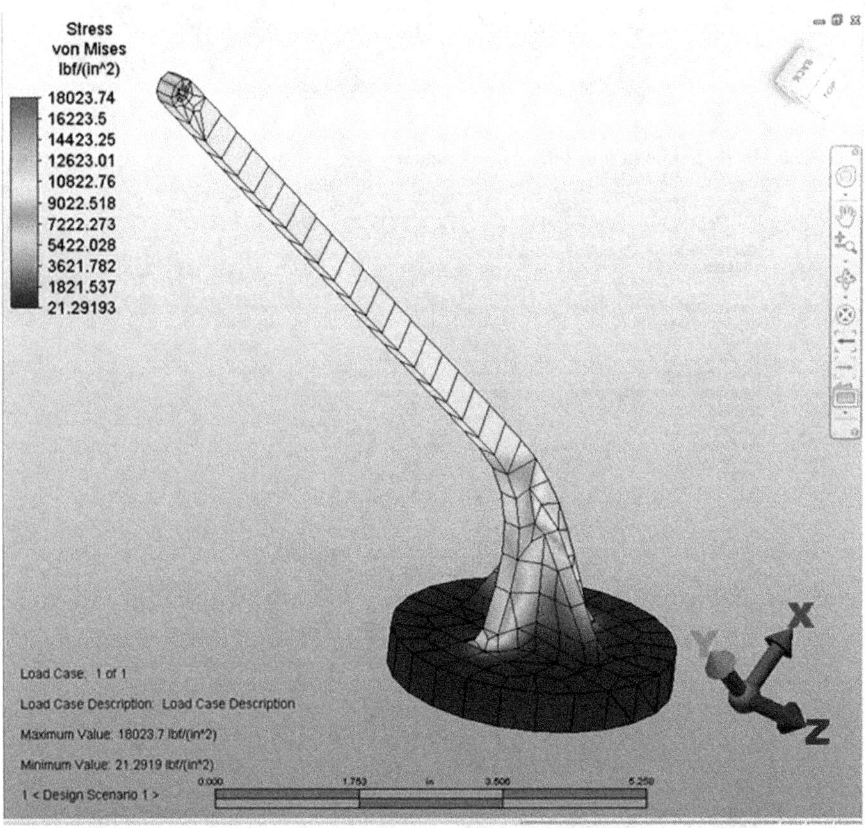

Fig. 4.28 Stress profile in orientation link

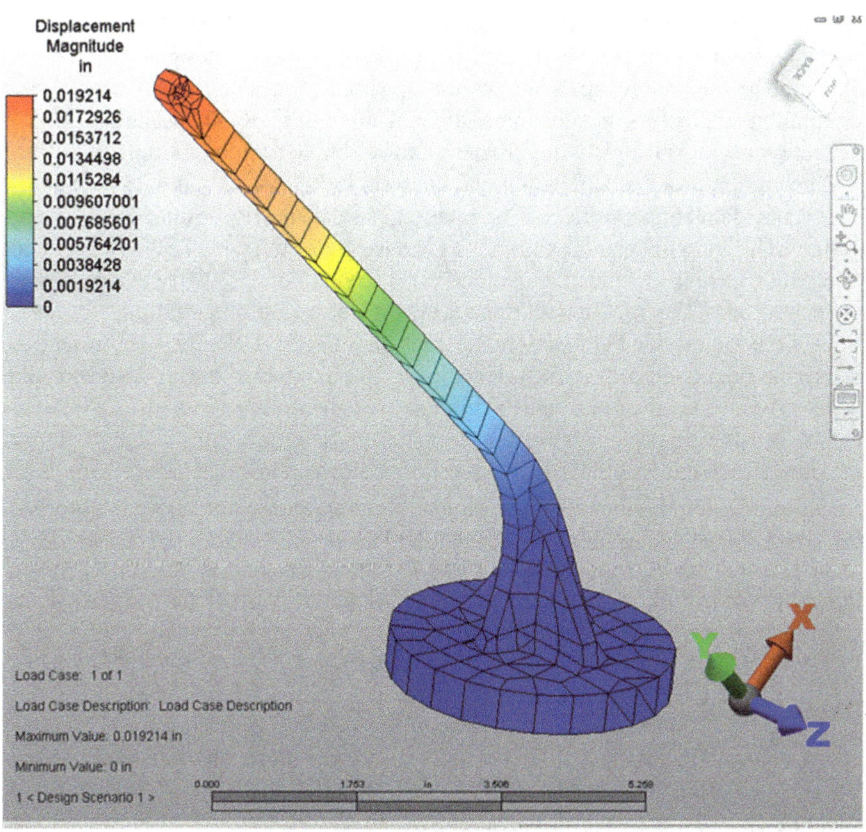

Fig. 4.29 Deflection profile in orientation link

The computer-aided simulation and analysis in Figs. 4.10 and 4.11 indicate the stress and deflection profiles of lateral channel beam in newly designed solar panel system. The analytic results state that the maximum stress of 15,877.88 psi in this lateral channel beam is less than the material yield strength of 36,300 psi and maximum deflection of 0.00828 in. is within material allowable deformation limit.

The computer-aided simulation and analysis in Figs. 4.13 and 4.14 display the stress and deflection profiles of longitudinal channel beam in newly designed solar panel system. The analytic results exhibit that the maximum stress of 17,589.58 psi in this longitudinal channel beam is less than the material yield strength of 36,300 psi and maximum deflection of 0.02748 in. is within material allowable deformation limit.

The computer-aided simulation and analysis in Figs. 4.16 and 4.17 present the stress and deflection profiles of orientation adaptor in newly designed solar panel system. The analytic results demonstrate that the maximum stress of 14,196.50 psi in this orientation adaptor is less than the material yield strength of 36,300 psi and maximum deflection of 0.00017 in. is within material allowable deformation limit.

The computer-aided simulation and analysis in Figs. 4.19 and 4.20 indicate the stress and deflection profiles of orientation rack in newly designed solar panel system. The analytic results state that the maximum stress of 16,842.35 psi in this orientation rack is less than the material yield strength of 36,300 psi and maximum deflection of 0.0511 in. is within material allowable deformation limit.

The computer-aided simulation and analysis in Figs. 4.22 and 4.23 display the stress and deflection profiles of orientation base in newly designed solar panel system. The analytic results exhibit that the maximum stress of 17,384.07 psi in this orientation base is less than the material yield strength of 36,300 psi and maximum deflection of 0.00345 in. is within material allowable deformation limit.

The computer-aided simulation and analysis in Figs. 4.25 and 4.26 present the stress and deflection profiles of orientation driver support in newly designed solar panel system. The analytic results exhibit that the maximum stress of 17,788.09 psi in this orientation base is less than the material yield strength of 36,300 psi and maximum deflection of 0.00016 in. is within material allowable deformation limit.

The computer-aided simulation and analysis in Figs. 4.28 and 4.29 demonstrate the stress and deflection profiles of orientation link in newly designed solar panel system. The analytic results indicate that the maximum stress of 18,023.74 psi in this orientation link is less than the material yield strength of 36,300 psi and maximum deflection of 0.01921 in. is within material allowable deformation limit.

Based on computer-aided analysis, shown in Figs. 4.3, 4.4, 4.5, 4.6, 4.7, 4.8, 4.9, 4.10, 4.11, 4.12, 4.13, 4.14, 4.15, 4.16, 4.17, 4.18, 4.19, 4.20, 4.21, 4.22, 4.23, 4.24, 4.25, 4.26, 4.27, 4.28, and 4.29, the maximum stresses generated in all important components including panel frame, lateral channel beam, longitudinal channel beam, orientation adaptor, orientation rack, orientation base, orientation driver support, and orientation link are all less than the material yield stress and the relevant maximum deflections of these components are all within allowable deformation limits of the materials. The above computational simulations have shown good performance of this newly developed solar tracking system.

4.3 Experiment on Solar Panel Tracking System

New solar panel tracking system has been prototyped and tested to compare and verify the computer-aided simulation results. Table 4.1 demonstrates the experimental results of maximum stress and maximum deflection of solar panel in this new solar panel tracking system.

The prototype experimental results of solar panel in Table 4.1 verify the proper function of this new solar panel system because the average maximum stress 2,267.160 psi and average maximum deflection 0.00519 in. are approximately equal to the results of maximum stress 2,267.076 psi and maximum deflection 0.00507 in. that are represented, respectively, in Figs. 4.4 and 4.5 by computer-aided simulation.

Table 4.2 expresses the experimental results of maximum stress and maximum deflection of solar panel frame in this new solar panel tracking system.

Table 4.1 Experimental results of maximum stress and maximum deflection of solar panel in this new solar panel tracking system

Number of experiment	Maximum stress (psi)	Maximum deflection (in.)
1	2,267.12	0.00505
2	2,267.09	0.00509
3	2,267.15	0.00508
4	2,267.06	0.00511
5	2,267.11	0.00515
6	2,267.15	0.00516
7	2,267.18	0.00524
8	2,267.19	0.00518
9	2,267.17	0.00519
10	2,267.15	0.00527
11	2,267.18	0.00525
12	2,267.21	0.00524
13	2,267.24	0.00522
14	2,267.22	0.00529
15	2,267.18	0.00527
16	2,267.14	0.00526
17	2,267.19	0.00529
18	2,267.18	0.00518
19	2,267.14	0.00519
20	2,267.17	0.00517
Average	2,267.16	0.00519

Table 4.2 Experimental results of maximum stress and maximum deflection of solar panel frame in this new solar panel tracking system

Number of experiment	Maximum stress (psi)	Maximum deflection (in.)
1	14,343.12	0.03124
2	14,343.11	0.03118
3	14,343.08	0.03119
4	14,343.09	0.03129
5	14,343.02	0.03124
6	14,343.15	0.03115
7	14,343.16	0.03148
8	14,343.18	0.03138
9	14,343.19	0.03131
10	14,343.14	0.03124
11	14,343.12	0.03122
12	14,343.01	0.03121
13	14,343.08	0.03118
14	14,343.17	0.03119
15	14,343.24	0.03117
16	14,343.29	0.03124
17	14,343.22	0.03148
18	14,343.24	0.03129
19	14,343.18	0.03125
20	14,343.16	0.03124
Average	14,343.15	0.03126

Table 4.3 Experimental results of maximum stress and maximum deflection of lateral channel beam in this new solar panel tracking system

Number of experiment	Maximum stress (psi)	Maximum deflection (in.)
1	15,877.78	0.00835
2	15,877.84	0.00838
3	15,877.69	0.00837
4	15,877.71	0.00836
5	15,877.68	0.00848
6	15,877.55	0.00846
7	15,877.68	0.00844
8	15,877.65	0.00849
9	15,877.59	0.00845
10	15,877.57	0.00842
11	15,877.66	0.00838
12	15,877.69	0.00847
13	15,877.68	0.00848
14	15,877.78	0.00837
15	15,877.77	0.00844
16	15,877.72	0.00845
17	15,877.67	0.00842
18	15,877.65	0.00848
19	15,877.68	0.00838
20	15,877.71	0.00841
Average	15,877.69	0.00842

The prototype experimental results of solar panel frame in Table 4.2 prove the normal function of solar panel frame as the average maximum stress 14,343.15 psi and average maximum deflection 0.03126 in. are almost same as the results of maximum stress 14,343.03 psi and maximum deflection 0.03143 in. that are indicated, respectively, in Figs. 4.7 and 4.8 by computer-aided simulation.

Table 4.3 records the experimental results of maximum stress and maximum deflection of lateral channel beam in this new solar panel tracking system.

The prototype experimental results of lateral channel beam in Table 4.3 confirm the appropriate function of lateral channel beam since the average maximum stress 15,877.69 psi and average maximum deflection 0.00842 in. are approximately same as the results of maximum stress 15,877.88 psi and maximum deflection 0.00827 in. that are laid out, respectively, in Figs. 4.10 and 4.11 by computer-aided simulation.

Table 4.4 states the experimental results of maximum stress and maximum deflection of longitudinal channel beam in this new solar panel tracking system.

The prototype experimental results of longitudinal channel beam in Table 4.4 verify proper function of this longitudinal channel beam because the average maximum stress 17,589.44 psi and average maximum deflection 0.02761 in. are very close to the results of maximum stress 17,589.58 psi and maximum deflection 0.02748 in. that are shown, respectively, in Figs. 4.13 and 4.14 by computer-aided simulation.

Table 4.4 Experimental results of maximum stress and maximum deflection of longitudinal channel beam in this new solar panel tracking system

Number of experiment	Maximum stress (psi)	Maximum deflection (in.)
1	17,589.48	0.02759
2	17,589.38	0.02754
3	17,589.39	0.02745
4	17,589.44	0.02759
5	17,589.47	0.02768
6	17,589.54	0.02766
7	17,589.49	0.02767
8	17,589.45	0.02765
9	17,589.42	0.02757
10	17,589.38	0.02759
11	17,589.47	0.02768
12	17,589.42	0.02769
13	17,589.48	0.02767
14	17,589.44	0.02768
15	17,589.45	0.02755
16	17,589.38	0.02763
17	17,589.39	0.02765
18	17,589.41	0.02761
19	17,589.42	0.02755
20	17,589.45	0.02759
Average	17,589.44	0.02761

Table 4.5 Experimental results of maximum stress and maximum deflection of orientation adaptor in this new solar panel tracking system

Number of experiment	Maximum stress (psi)	Maximum deflection (in.)
1	14,196.45	0.00019
2	14,196.51	0.00016
3	14,196.42	0.00012
4	14,196.41	0.00014
5	14,196.48	0.00011
6	14,196.35	0.00012
7	14,196.38	0.00011
8	14,196.37	0.00013
9	14,196.31	0.00012
10	14,196.33	0.00011
11	14,196.38	0.00010
12	14,196.41	0.00008
13	14,196.42	0.00009
14	14,196.36	0.00011
15	14,196.35	0.00008
16	14,196.32	0.00007
17	14,196.31	0.00011
18	14,196.33	0.00010
19	14,196.35	0.00012
20	14,196.36	0.00010
Average	14,196.38	0.00011

Table 4.6 Experimental results of maximum stress and maximum deflection of orientation rack in this new solar panel tracking system

Number of experiment	Maximum stress (psi)	Maximum deflection (in.)
1	16,842.25	0.05109
2	16,842.24	0.05115
3	16,842.31	0.05114
4	16,842.21	0.05118
5	16,842.23	0.05117
6	16,842.22	0.05119
7	16,842.21	0.05117
8	16,842.24	0.05115
9	16,842.18	0.05124
10	16,842.19	0.05118
11	16,842.20	0.05115
12	16,842.24	0.05117
13	16,842.21	0.05118
14	16,842.19	0.05119
15	16,842.21	0.05124
16	16,842.18	0.05121
17	16,842.22	0.05122
18	16,842.24	0.05117
19	16,842.18	0.05118
20	16,842.24	0.05116
Average	16,842.22	0.05118

Table 4.5 presents the experimental results of maximum stress and maximum deflection of orientation adaptor in this new solar panel tracking system.

The prototype experimental results of orientation adaptor in Table 4.5 prove the normal function of orientation adaptor as the average maximum stress 14,196.38 psi and average maximum deflection 0.00011 in. are similar to the results of maximum stress 14,196.50 psi and maximum deflection 0.00017 in. that are presented, respectively, in Figs. 4.16 and 4.17 by computer-aided simulation.

Table 4.6 expresses the experimental results of maximum stress and maximum deflection of orientation rack in this new solar panel tracking system.

The prototype experimental results of orientation rack in Table 4.6 confirm the appropriate function of orientation rack since the average maximum stress 16,842.22 psi and average maximum deflection 0.05118 in. are almost equal to the results of maximum stress 16,842.35 psi and maximum deflection 0.05107 in. that are indicated, respectively, in Figs. 4.19 and 4.20 by computer-aided simulation.

Table 4.7 conveys the experimental results of maximum stress and maximum deflection of orientation base in this new solar panel tracking system.

The prototype experimental results of orientation base in Table 4.7 prove the normal function of orientation base as the average maximum stress 17,384.20 psi and average maximum deflection 0.00331 in. are closely equal to the results of

Table 4.7 Experimental results of maximum stress and maximum deflection of orientation base in this new solar panel tracking system

Number of experiment	Maximum stress (psi)	Maximum deflection (in.)
1	17,384.15	0.00338
2	17,384.17	0.00333
3	17,384.06	0.00348
4	17,384.18	0.00341
5	17,384.24	0.00337
6	17,384.22	0.00324
7	17,384.25	0.00329
8	17,384.21	0.00327
9	17,384.17	0.00325
10	17,384.18	0.00332
11	17,384.19	0.00327
12	17,384.18	0.00325
13	17,384.22	0.00327
14	17,384.24	0.00330
15	17,384.25	0.00327
16	17,384.26	0.00329
17	17,384.24	0.00324
18	17,384.19	0.00326
19	17,384.18	0.00331
20	17,384.17	0.00330
Average	17,384.20	0.00331

Table 4.8 Experimental results of maximum stress and maximum deflection of orientation driver support in this new solar panel tracking system

Number of experiment	Maximum stress (psi)	Maximum deflection (in.)
1	17,788.15	0.00017
2	17,788.08	0.00019
3	17,788.17	0.00018
4	17,788.18	0.00024
5	17,788.19	0.00025
6	17,788.16	0.00026
7	17,788.08	0.00023
8	17,788.12	0.00025
9	17,788.19	0.00029
10	17,788.18	0.00027
11	17,788.24	0.00029
12	17,788.24	0.00027
13	17,788.21	0.00024
14	17,788.27	0.00026
15	17,788.26	0.00025
16	17,788.27	0.00024
17	17,788.24	0.00029
18	17,788.22	0.00027
19	17,788.24	0.00026
20	17,788.21	0.00024
Average	17,788.20	0.00025

Table 4.9 Experimental results of maximum stress and maximum deflection of orientation link in this new solar panel tracking system

Number of experiment	Maximum stress (psi)	Maximum deflection (in.)
1	18,023.68	0.01915
2	18,023.66	0.01916
3	18,023.62	0.01918
4	18,023.54	0.01911
5	18,023.59	0.01912
6	18,023.78	0.01911
7	18,023.67	0.01915
8	18,023.62	0.01910
9	18,023.60	0.01911
10	18,023.61	0.01912
11	18,023.54	0.01908
12	18,023.55	0.01910
13	18,023.57	0.01911
14	18,023.56	0.01912
15	18,023.57	0.01914
16	18,023.55	0.01915
17	18,023.59	0.01911
18	18,023.61	0.01912
19	18,023.62	0.01910
20	18,023.63	0.01914
Average	18,023.61	0.01912

maximum stress 17,384.07 psi and maximum deflection 0.00335 in. that are recorded, respectively, in Figs. 4.22 and 4.23 by computer-aided simulation.

Table 4.8 lays out the experimental results of maximum stress and maximum deflection of orientation driver support in this new solar panel tracking system.

The prototype experimental results of orientation driver support in Table 4.8 verify the proper function of orientation driver support because the average maximum stress 17,788.20 psi and average maximum deflection 0.00025 in. are almost same as the results of maximum stress 17,788.09 psi and maximum deflection 0.00016 in. that are displayed, respectively, in Figs. 4.25 and 4.26 by computer-aided simulation.

Table 4.9 shows the experimental results of maximum stress and maximum deflection of orientation link in this new solar panel tracking system.

The prototype experimental results of orientation link in Table 4.9 confirm the appropriate function of orientation link since the average maximum stress 18,023.61 psi and average maximum deflection 0.01912 in. are approximately equal to the results of maximum stress 18,023.74 psi and maximum deflection 0.01921 in. that are exhibited, respectively, in Figs. 4.28 and 4.29 by computer-aided simulation.

4.4 Discussion and Future Improvement on Solar Panel Tracking System

Solar power system can transfer the natural sunlight energy to the electrical mechanical energies. It has the major advantages including safe and clean features compared to the traditional energy resources. This chapter introduces the methodologies using computational simulation and prototype testing to determine the important factors affecting the performance of this new solar panel tracking system. Both computational simulation and prototype experiment show its reliable function. Future improvement has been planned to optimize this solar tracking system by computer-aided modeling and analysis to determine the optimal working condition, further reduce system weight to save system cost, and revise the gear train for more quiet operation.

Energy-Saving Cooling System 5

5.1 Design of Energy-Saving Cooling System

The energy-saving cooling systems have been investigated in recent years and more R&D efforts have been put to improve its performance with higher energy-saving features. This new automatically controlled energy-saving cooling system has been developed, analyzed, and prototyped to verify its function and reliability ability. The major advantages of this new system include simpler structure, compact setup, and easy parameter adjustment to improve its performance, increased efficiency, quiet operation, and less frictional losses in this free piston cooling system. The gaseous pressure inside system can be manipulated by two automatic valves controlled through PLC program in the controller and the motion of free piston is monitored by proximity sensors to improve the performance by automatically adjusting the functioning parameters. The prototype and internal view are shown in Figs. 5.1 and 5.2.

In this new design, partial compressive work in the middle fluctuating chamber can be transferred into the useful motive work if compared with some cooling systems such as Solvay system in which no energy can be retrieved from middle chamber and full compressive work is transformed into the heat discharged to the air. Furthermore, the functioning efficiency in this system is improved because the cooling capacity can be output at the two ends of systematic cylinder. This new system operates quietly since the vibration and shock are decreased due to its symmetrical setup design. The systematic phase angle of various thermal and mechanical function curves can be easily manipulated and adjusted through automatically controlled valves to improve its function and efficiency. Figures 5.1 and 5.2 display the external and internal view of this newly developed cooling system. Two proximity sensors, two compressive cylinders V_c (V'_c), and two expansive cylinders V_e (V'_e) are symmetrically installed in the cooling system. Two free pistons F (F′), one thin plate G mounted in the middle of cylinder with a

J. Zheng Li, *CAD, 3D Modeling, Engineering Analysis, and Prototype Experimentation*, DOI 10.1007/978-3-319-05921-1_5

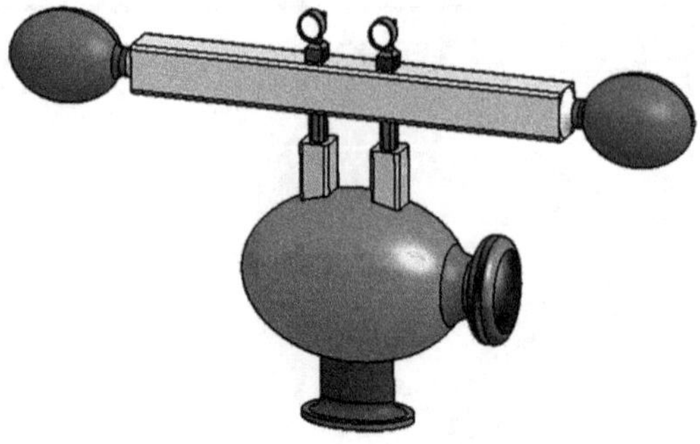

Fig. 5.1 New energy-saving cooling system

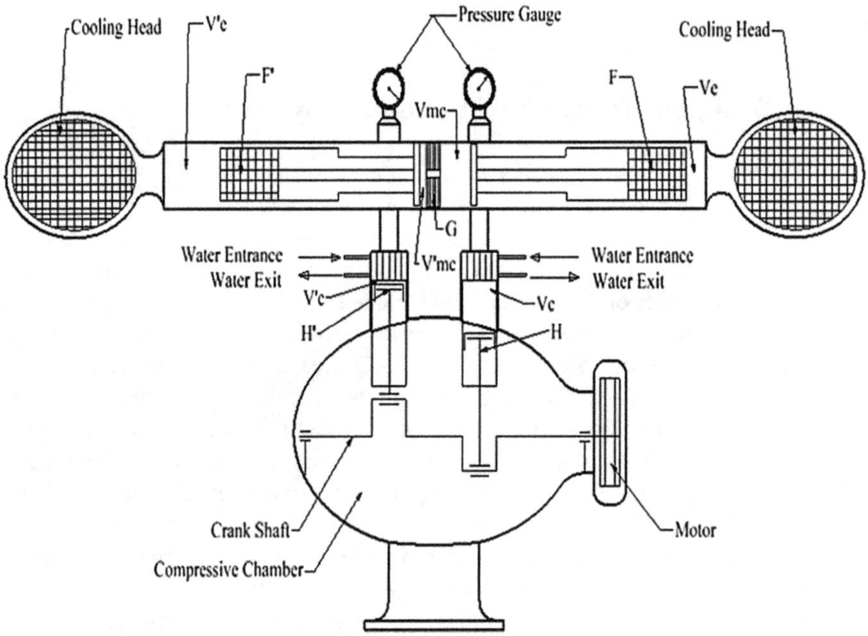

Fig. 5.2 Internal view of new energy-saving cooling system

small hole on it, and two middle chambers V_{mc} (V'_{mc}) are set up in this system. Due to its symmetrical design, only chambers of V_c, V_e, and V_{mc}, compressor piston H, and free piston F in this system are required to be analyzed. With the installation of two automatically controlled valves, the air pressure in V_{mc} and V'_{mc} can be changed to adjust the phase angles of all functional curves to improve the machine function through PLC-controlled automatic valves. Two electric sensors at the two ends of the cylinder can monitor the piston's moving distance that can be fed back to the PLC unit to readjust the thermal parameters to improve the system function.

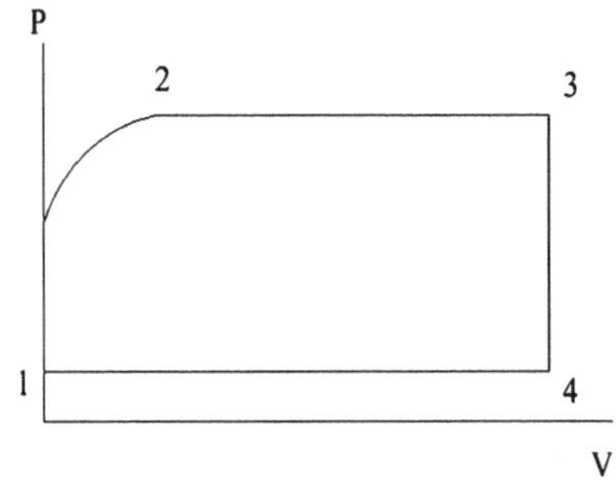

Fig. 5.3 *P–V* cyclic diagram in expansive chamber

5.2 Computer-Aided Simulation on Energy-Saving Cooling System

The initial cyclic condition of this new energy-saving cooling system is previously displayed in Fig. 5.2. In this stable working condition, piston H locates at the lower position in the compressive chamber and free piston F is at the right end of the expansive chamber. The *P–V* cyclic diagram of expansive chamber is shown in Fig. 5.3.

Because the gaseous media with high pressure ($P = P_H$) enters chamber V_{mc} via a tiny hole at the end time period of last functioning cycle, gaseous media with low pressure ($P = P_L$) in chamber V_{mc} will be mixed with high-pressure gaseous media. The mixed gaseous pressure in chamber V_{mc} is P_{mix}. When piston H starts moving upward in compressive chamber V_c, gaseous media in chamber V_{mc} is assumed not entering chamber V'_{mc} since the traveling speed of free piston is fast and the diameter of hole is very small. Because the left-side gaseous pressure of free piston F is higher than the right-side gaseous pressure of F ($P_{mc} = P_{ave} > P_e$), free piston F does not move and the gaseous pressure P_e in expansive chamber reaches high pressure P_H and gaseous volume V_e in expansive chamber increases from zero to V_2 as piston H continues moving up. This process can be represented by curve 1–2 in Fig. 5.3. Then the gaseous media in chamber V_{mc} flows into chamber V'_{mc}. Since the gaseous pressure at the left and right sides of free piston F is same, F travels to the left with constant speed. When crankshaft rotates to the angle (θ) near 180°, the gaseous volume V_e in expansive chamber increases from V_2 to V_3 which can be represented by the line 2–3. During initial period that piston H moves downwards in chamber V_c, free piston F does not move since the gaseous media in chamber V'_{mc} is assumed not to flow into chamber V_{mc}, and pressure P_e in expansive chamber reduces from P_3 to P_4 which is represented by line 3–1. When piston H continues traveling downwards in compressive chamber V_c, gaseous media in chamber V'_{mc} enters chamber V_{mc} via tiny hole and free piston F moves to the right end in the expansive chamber with constant speed which can be represented by line 4–1. This concludes the full functioning cycle of this cooling system.

The computer-aided simulations can be applied to verify the function of this new cooling system. The piston H in compressive chamber moves in sine law (Kundu and Cohen 2008):

$$V_{\mathrm{c}}(\theta)=\frac{M\times V_{\mathrm{co}}\times[1+\cos(\theta)]}{2} \tag{5.1}$$

$$V=V_{\mathrm{k}}+V_{\mathrm{c}}+V_{\mathrm{e}} \tag{5.2}$$

Here, V_{k} is the clearance volume:

$$P=R\times\sum\frac{M_{\mathrm{i}}}{\frac{V_{\mathrm{k}}}{T_{\mathrm{k}}}}+\frac{V_{\mathrm{e}}}{T_{\mathrm{e}}}+\frac{V_{\mathrm{c}}}{T_{\mathrm{c}}} \tag{5.3}$$

As piston H in compressive chamber travels upwards with crankshaft rotating to angle ω,

$$(P_{\mathrm{e}}-P_{\mathrm{me}})\times A=F_{\mathrm{F}} \tag{5.4}$$

$$P(\omega)=R\times\sum\frac{M_{\mathrm{i}}}{\frac{V_{\mathrm{k}}}{T_{\mathrm{k}}}}+0+\frac{V_{\mathrm{c}}(\omega)}{T_{\mathrm{c}}} \tag{5.5}$$

Assuming $W=\frac{T_{\mathrm{c}}}{T_{\mathrm{k}}}$, $L=\frac{V_{\mathrm{k}}}{V_{\mathrm{eo}}}$, and $Q=\frac{T_{\mathrm{c}}}{T_{\mathrm{e}}}$

$$P=\frac{P(\omega)\times\left[L\times W+\frac{V_{\mathrm{c}}(\omega)}{V_{\mathrm{eo}}}\right]}{\left[L\times W+\frac{V_{\mathrm{e}}}{V_{\mathrm{eo}}}\right]}\times Q+\frac{V_{\mathrm{c}}(\omega)}{V_{\mathrm{eo}}} \tag{5.6}$$

The differential equation of free piston motion can be derived based on the second Newton law:

$$m\times\left[\frac{\mathrm{d}^2(V_{\mathrm{e}}(\theta)}{\mathrm{d}t^2}\right]=\left[P_{\mathrm{e}}-P_{\mathrm{mc}}\right)\times S]-F_{\mathrm{F}} \tag{5.7}$$

Combine the above equations:

$$\frac{\mathrm{d}^2\left[\frac{V_{\mathrm{e}}(\theta)}{V_{\mathrm{eo}}}\right]}{\mathrm{d}\theta^2}=\frac{P(\omega)\times\left[L\times W+\frac{V_{\mathrm{c}}(\omega)}{V_{\mathrm{eo}}}\right]}{\frac{m\times w^2\times Y_{\mathrm{o}}}{A}}$$

$$\frac{1}{L\times W+\frac{Q\times V_{\mathrm{e}}(\theta)}{V_{\mathrm{eo}}}+\frac{V_{\mathrm{c}}(\theta)}{V_{\mathrm{eo}}}\times\frac{A\times\left(P_{\mathrm{L}}+P_{H}\right)}{2\times m\times w^2\times Y_{\mathrm{o}}}} \tag{5.8}$$

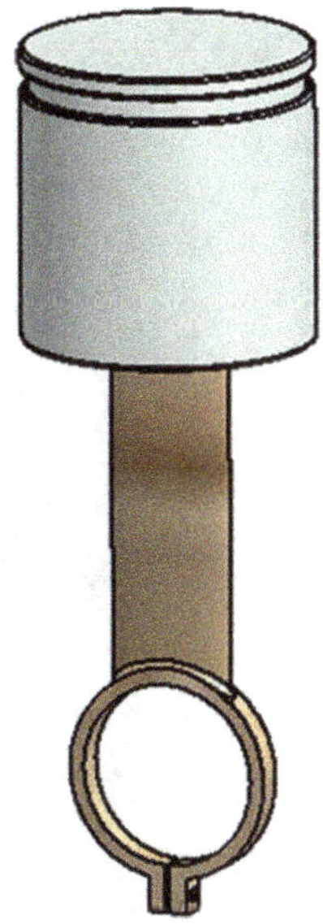

Fig. 5.4 Piston and link assembly in compressive chamber

Here, pressure $P(\omega)$ can be determined by the following equation:

$$\frac{d^2 V_c(\theta)}{d\theta^2} = 0 \tag{5.9}$$

The computer-aided simulation determines the stress and deflection profiles shown in Figs. 5.4, 5.5, 5.6, 5.7, 5.8, 5.9, 5.10, 5.11, 5.12, 5.13, 5.14, 5.15, 5.16, 5.17, and 5.18.

The computer-aided simulation and analysis in Figs. 5.5 and 5.6 show the stress and deflection profiles of compressive piston link assembly in newly designed energy-saving cooling system. The analytic results exhibit that the maximum stress of 25,698.25 psi in this piston link assembly is less than the material yield strength of 36,300 psi and maximum deflection of 0.00169 in. is within material allowable deformation limit.

The computer-aided simulation and analysis in Figs. 5.8 and 5.9 present the stress and deflection profiles of compressive chamber in newly designed energy-saving cooling system. The analytic results demonstrate that the maximum stress of 26,088.02 psi in this compressive chamber is less than the material yield strength of 36,300 psi and maximum deflection of 0.00017 in. is within material allowable deformation limit.

The computer-aided simulation and analysis in Figs. 5.11 and 5.12 indicate the stress and deflection profiles of piston link in newly designed energy-saving cooling system. The analytic results state that the maximum stress of 23,242.99 psi in this piston link is less than the material yield strength of 36,300 psi and maximum deflection of 0.00147 in. is within material allowable deformation limit.

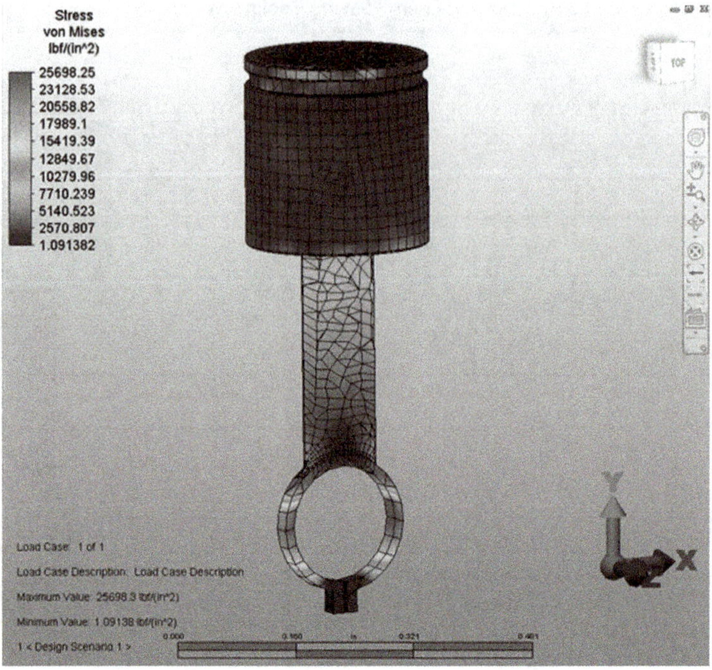

Fig. 5.5 Stress profile of compressive piston link assembly

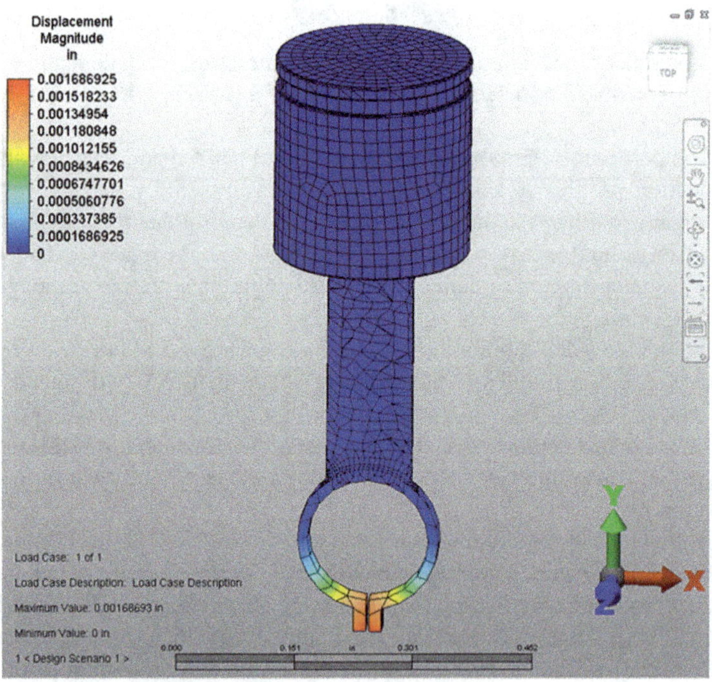

Fig. 5.6 Deflection profile of compressive piston link assembly

Fig. 5.7 Piston head in compressive chamber

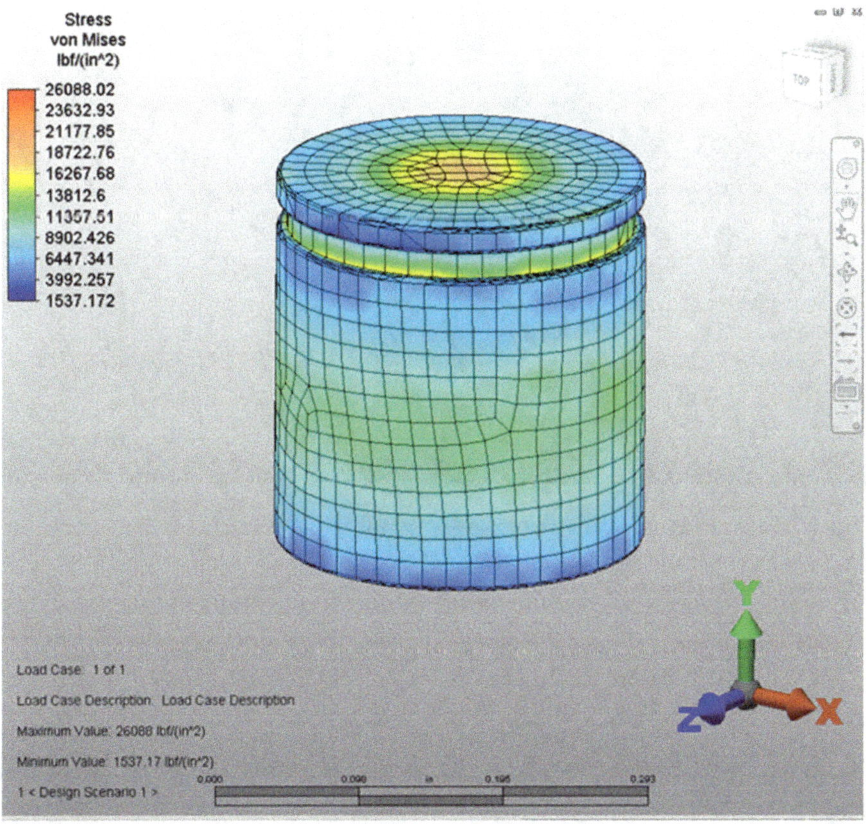

Fig. 5.8 Stress profile of piston head in compressive chamber

The computer-aided simulation and analysis in Figs. 5.14 and 5.15 display the stress and deflection profiles of crankshaft in newly designed energy-saving cooling system. The analytic results show that the maximum stress of 20,667.27 psi in this crankshaft is less than the material yield strength of 36,300 psi and maximum deflection of 0.00019 in. is within material allowable deformation limit.

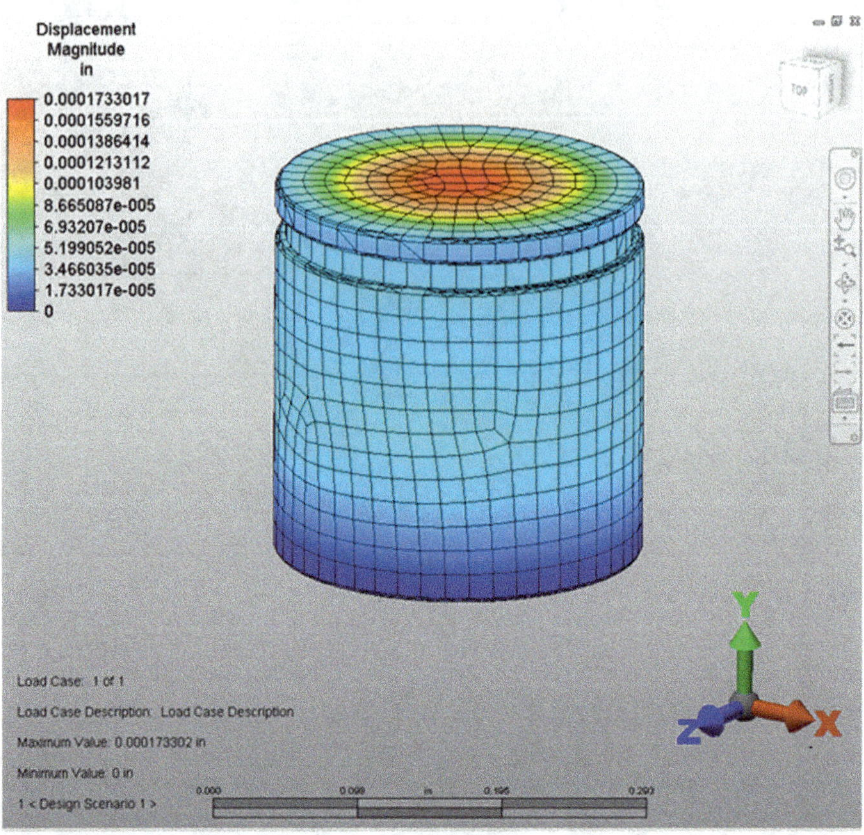

Fig. 5.9 Deflection profile of piston head in compressive chamber

Fig. 5.10 Piston link in compressive chamber

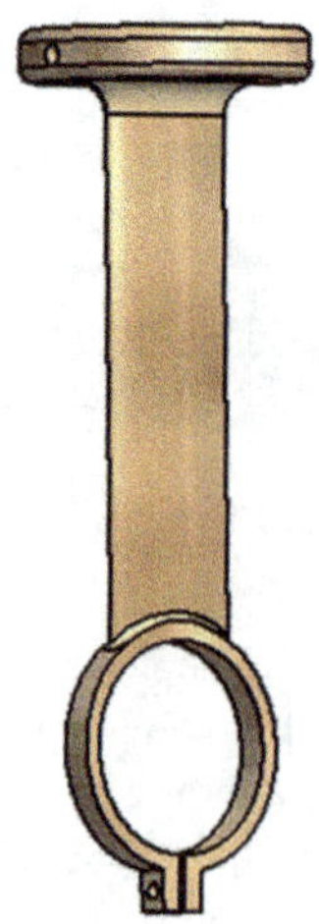

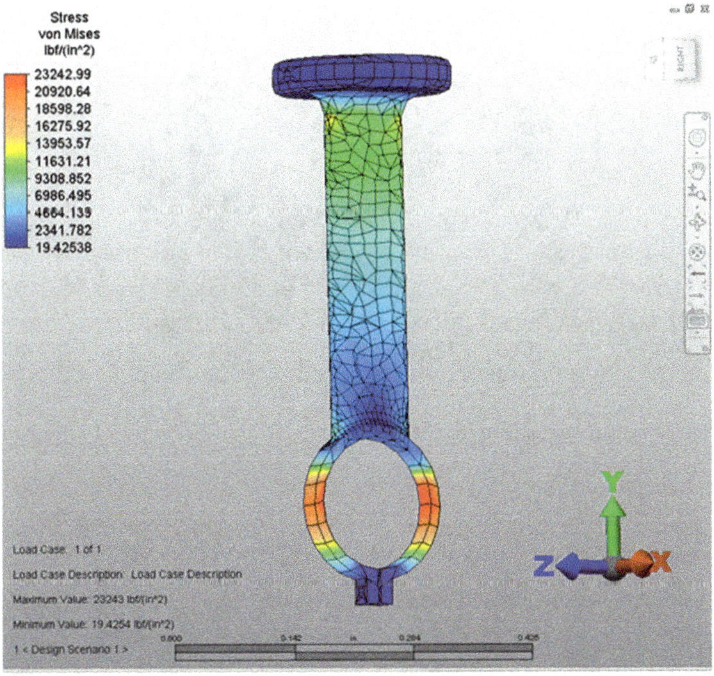

Fig. 5.11 Stress profile of piston link in compressive chamber

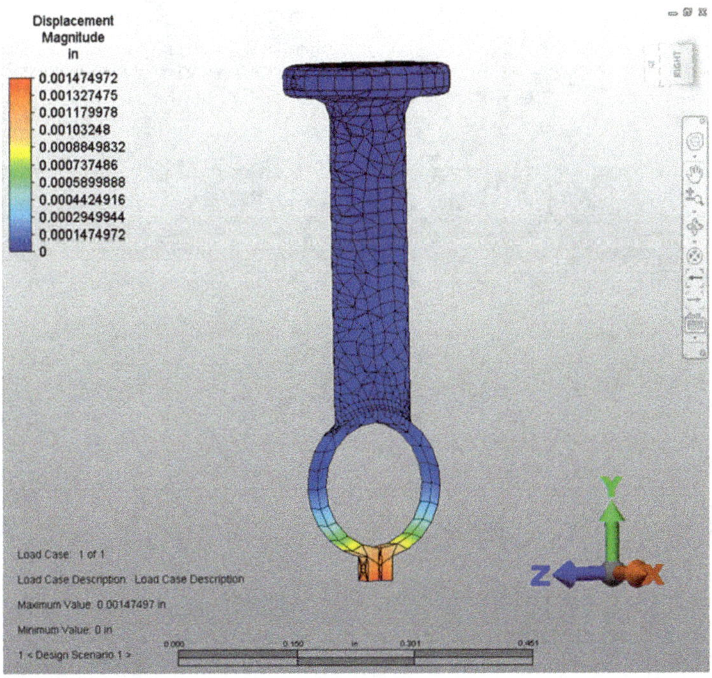

Fig. 5.12 Deflection profile of piston link in compressive chamber

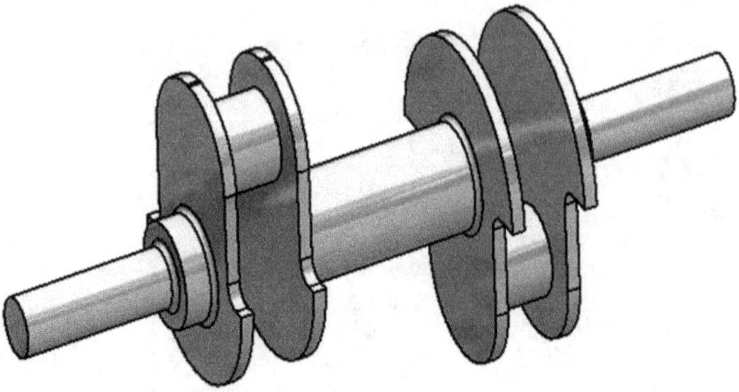

Fig. 5.13 Crankshaft in compressive chamber

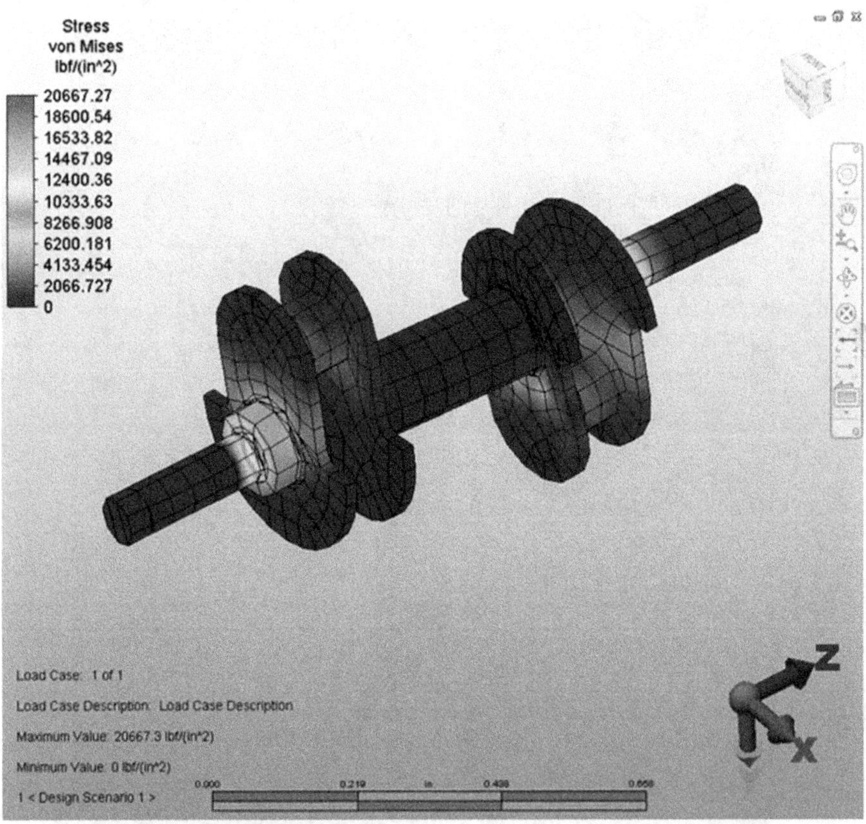

Fig. 5.14 Stress profile of crankshaft in compressive chamber

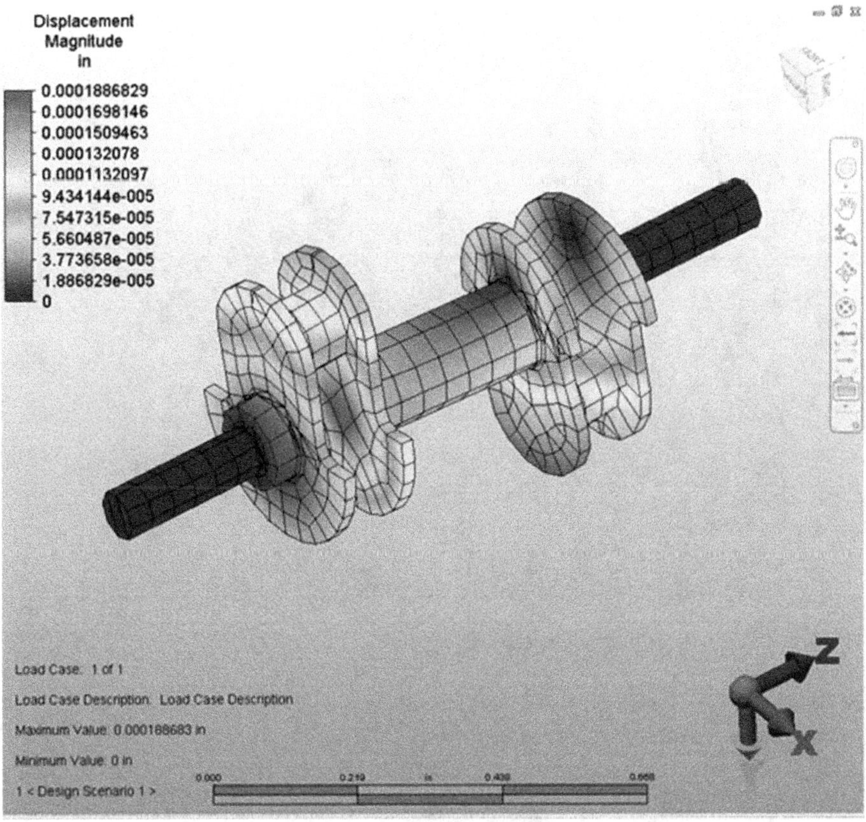

Fig. 5.15 Deflection profile of crankshaft in compressive chamber

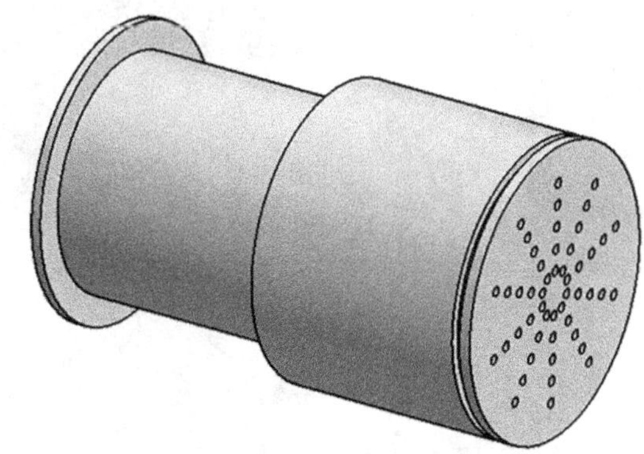

Fig. 5.16 Piston in expansive chamber

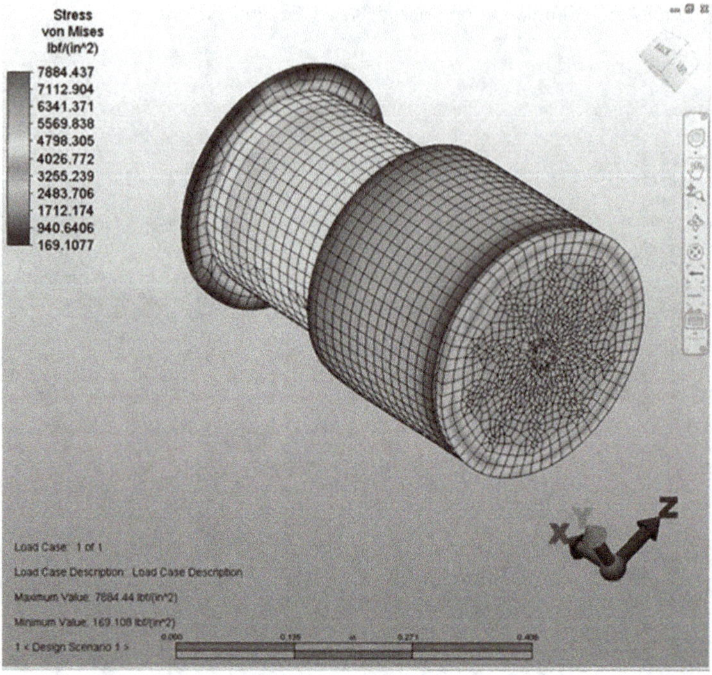

Fig. 5.17 Stress profile of piston in expansive chamber

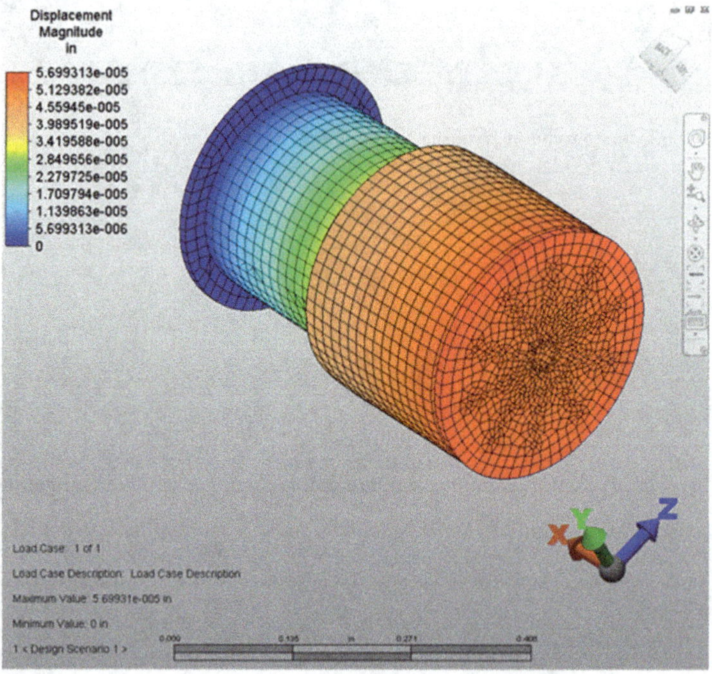

Fig. 5.18 Deflection profile of piston in expansive chamber

The computer-aided simulation and analysis in Figs. 5.17 and 5.18 exhibit the stress and deflection profiles of piston in newly designed energy-saving cooling system. The analytic results present that the maximum stress of 7,884.44 psi in this piston is less than the material yield strength of 36,300 psi and maximum deflection of 0.00006 in. is within material allowable deformation limit.

The above computational simulation results demonstrate that the maximum stresses on these critical components are all less than the material yield stress and maximum material deflections are all within material allowable deformation limits. The computational solutions confirm the good and reliable function of this newly developed energy-saving cooling system.

5.3 Experiment on Energy-Saving Cooling System

The newly designed energy-saving cooling system has been prototyped and tested to compare and verify the results from computer-aided simulation. Table 5.1 demonstrates the prototype testing results of compressive piston link assembly in this new energy-saving cooling system.

The prototype experimental results of compressive piston link assembly in Table 5.1 verify the proper function of this compressive piston link assembly

Table 5.1 Prototype testing of compressive piston link assembly in this new energy-saving cooling system

Number of experiment	Maximum stress (psi)	Maximum deflection (in.)
1	25,698.33	0.00159
2	25,698.29	0.00161
3	25,698.38	0.00155
4	25,698.45	0.00148
5	25,698.49	0.00139
6	25,698.48	0.00138
7	25,698.55	0.00144
8	25,698.49	0.00138
9	25,698.57	0.00147
10	25,698.54	0.00139
11	25,698.56	0.00138
12	25,698.53	0.00154
13	25,698.54	0.00149
14	25,698.56	0.00142
15	25,698.59	0.00136
16	25,698.57	0.00138
17	25,698.56	0.00135
18	25,698.55	0.00145
19	25,698.48	0.00144
20	25,698.42	0.00148
Average	25,698.50	0.00145

Table 5.2 Prototype testing of piston head of compressor chamber in this new energy-saving cooling system

Number of experiment	Maximum stress (psi)	Maximum deflection (in.)
1	26,088.12	0.00018
2	26,088.17	0.00019
3	26,088.11	0.00023
4	26,088.24	0.00019
5	26,088.29	0.00024
6	26,088.33	0.00026
7	26,088.38	0.00029
8	26,088.39	0.00032
9	26,088.35	0.00031
10	26,088.21	0.00036
11	26,088.18	0.00035
12	26,088.24	0.00033
13	26,088.35	0.00029
14	26,088.38	0.00027
15	26,088.39	0.00025
16	26,088.37	0.00022
17	26,088.33	0.00019
18	26,088.29	0.00021
19	26,088.27	0.00023
20	26,088.26	0.00022
Average	26,088.28	0.00026

because the average maximum stress 25,698.50 psi and average maximum deflection 0.00145 in. are approximately equal to the results of maximum stress 25,698.25 psi and maximum deflection 0.00169 in. that are represented, respectively, in Figs. 5.5 and 5.6 by computer-aided modeling and numerical simulation.

Table 5.2 expresses the prototype testing results of piston head of compressor chamber in this new energy-saving cooling system.

The prototype experimental results for piston head of compressor chamber in Table 5.2 confirm the appropriate function of piston head since the average maximum stress 26,088.28 psi and average maximum deflection 0.00026 in. are approximately same as the results of maximum stress 26,088.02 psi and maximum deflection 0.00017 in. that are indicated, respectively, in Figs. 5.8 and 5.9 by computer-aided modeling and numerical simulation.

Table 5.3 records the prototype testing results of piston link of compressor chamber in this new energy-saving cooling system.

The prototype experimental results for piston link of compressor chamber in Table 5.3 prove the normal function as the average maximum stress 23,242.65 psi and average maximum deflection 0.00175 in. are close to the results of maximum stress 23,242.99 psi and maximum deflection 0.00147 in. that are laid out, respectively, in Figs. 5.11 and 5.12 by computer-aided modeling and numerical simulation.

Table 5.3 Prototype testing of piston link of compressor chamber in this new energy-saving cooling system

Number of experiment	Maximum stress (psi)	Maximum deflection (in.)
1	23,242.78	0.00148
2	23,242.92	0.00155
3	23,242.88	0.00167
4	23,242.56	0.00178
5	23,242.48	0.00167
6	23,242.68	0.00179
7	23,242.85	0.00188
8	23,242.84	0.00187
9	23,242.68	0.00178
10	23,242.55	0.00184
11	23,242.68	0.00186
12	23,242.54	0.00179
13	23,242.69	0.00178
14	23,242.48	0.00168
15	23,242.49	0.00172
16	23,242.51	0.00184
17	23,242.38	0.00178
18	23,242.47	0.00177
19	23,242.84	0.00175
20	23,242.68	0.00176
Average	23,242.65	0.00175

Table 5.4 states the prototype testing results of crankshaft of compressor chamber in this new energy-saving cooling system.

The prototype experimental results for crankshaft of compressor chamber in Table 5.4 verify the proper function of crankshaft because the average maximum stress 20,667.43 psi and average maximum deflection 0.00013 in. are similar to the results of maximum stress 20,667.27 psi and maximum deflection 0.00019 in. that are shown, respectively, in Figs. 5.14 and 5.15 by computer-aided modeling and numerical simulation.

Table 5.5 demonstrates the prototype testing results of piston in compressor chamber of this new energy-saving cooling system.

The prototype experimental results of piston in compressor chamber in Table 5.5 confirm the appropriate function of this piston since the average maximum stress 7,884.67 psi and average maximum deflection 0.00010 in. are almost equal to the results of maximum stress 7,884.44 psi and maximum deflection 0.00006 in. that are presented, respectively, in Figs. 5.17 and 5.18 by computer-aided modeling and numerical simulation.

Table 5.4 Prototype testing of crankshaft of compressor chamber in this new energy-saving cooling system

Number of experiment	Maximum stress (psi)	Maximum deflection (in.)
1	20,667.35	0.00017
2	20,667.29	0.00015
3	20,667.38	0.00022
4	20,667.24	0.00015
5	20,667.39	0.00012
6	20,667.45	0.00014
7	20,667.48	0.00011
8	20,667.55	0.00012
9	20,667.59	0.00018
10	20,667.56	0.00011
11	20,667.52	0.00012
12	20,667.54	0.00014
13	20,667.52	0.00011
14	20,667.48	0.00009
15	20,667.45	0.00008
16	20,667.38	0.00009
17	20,667.35	0.00012
18	20,667.24	0.00008
19	20,667.37	0.00011
20	20,667.44	0.00012
Average	20,667.43	0.00013

Table 5.5 Prototype testing of piston in compressor chamber of this new energy-saving cooling system

Number of experiment	Maximum stress (psi)	Maximum deflection (in.)
1	7,884.54	0.00008
2	7,884.41	0.00011
3	7,884.55	0.00012
4	7,884.84	0.00009
5	7,884.88	0.00011
6	7,884.78	0.00005
7	7,884.75	0.00012
8	7,884.69	0.00009
9	7,884.68	0.00008
10	7,884.54	0.00009
11	7,884.55	0.00011
12	7,884.42	0.00012
13	7,884.41	0.00013
14	7,884.59	0.00012
15	7,884.75	0.00011
16	7,884.84	0.00012
17	7,884.89	0.00013
18	7,884.72	0.00011
19	7,884.88	0.00008
20	7,884.62	0.00009
Average	7,884.67	0.00010

5.4 Discussion and Future Improvement on Energy-Saving Cooling System

The environment protection is one of the most important ways against global warming. The demand for saving energy and improving performance in cooling system helps environmental conservation and customer satisfaction. The systematic efficiency improvement is indispensable for saving energy in cooling system. This new cooling system shows its reliable function with good energy-saving and efficiency-increased features. The future improvement will include symmetrically balancing the piston setup featuring reduced shock, lower vibration, quiet operation, and higher efficiency in cooling system. Some design simplification, such as free piston glove cutoff, heat exchanger, and compressive crankshaft, will be planned for further cost reduction.

Part II

Automated Systems

Automated and High-Speed Manufacturing System

6

6.1 Design of Automated and High-Speed Manufacturing Systems

Automated production and manufacturing are the engineering fields dealing with various product manufacturing, production processes, machining equipment developments, and integration of production systems and manufacturing equipment. Computer-aided engineering (CAE) technology can provide engineering design teams with quick and cost-effective professional tools to efficiently iterate the design process for better quality and reliable function. The CAE technique can allow multiple design concepts being reviewed and evaluated without real product manufactured until the design process has been completed. Computer-aided manufacturing (CAM) helps to ensure qualified production process by computer-integrated technology that permits the production processes interchanging technical data information to each other. Figures 6.1, 6.2, and 6.3 show the newly developed automated/high-speed system to fill the high-viscous liquids, based on author's current research.

This automated and high-speed liquid filling system is designed to fill different high-viscous liquids. It can be flexibly applied to many industries including chemical, pharmaceutical, dairy, cosmetic, and food production. Some very thick liquids including medical cream, cosmetic products, and food sauces can be filled into the bottles and containers using this new automated filling system with positive displacement pump applied in the heavy viscous liquids under high-temperature environment and rotary gear pump in the heavy duty work for filling oil products, construction tar, roofing bitumen, thick ink, and special wax. Specially designed jacket inside rotary gear pump can keep pump working at raised temperature up to 125 °C and double-drive rotors in the pump make this system efficiently delivering high-viscosity liquids. To accelerate the liquid filling speed, multiple pumping nozzles with various sizes can be integrated in the filling system. The internally swaged nozzles are applied in this new automated liquid filling system for the bottles with narrow bottlenecks and complex geometrical shapes. The well-

J. Zheng Li, *CAD, 3D Modeling, Engineering Analysis, and Prototype Experimentation*, DOI 10.1007/978-3-319-05921-1_6

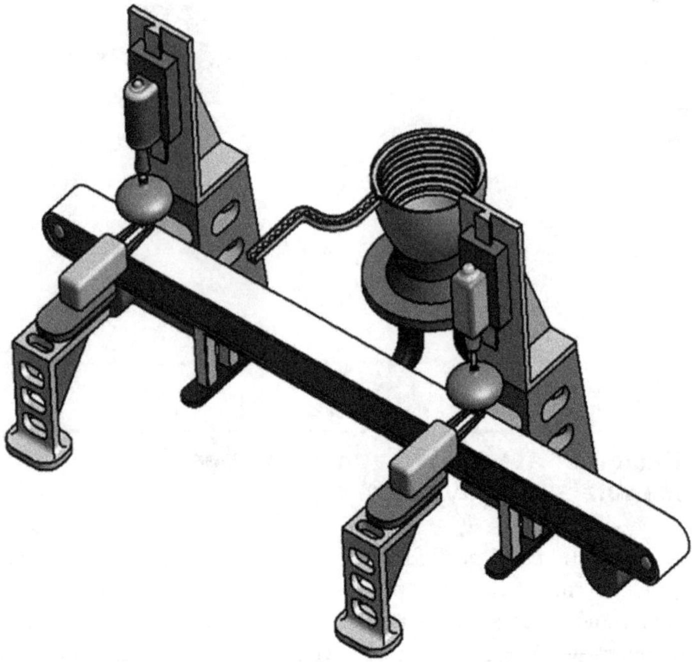

Fig. 6.1 Automated high-viscous liquid filling and sealing manufacturing system

Fig. 6.2 Automated high-viscous liquid filling mechanism

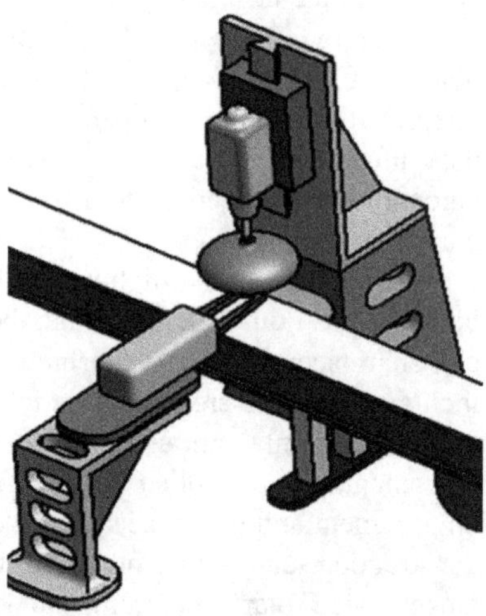

designed indexing conveyor controls the product movement within targeted tolerance range and empty bottles or containers are picked up and placed to the fixture holders secured in the indexing conveyer by specially designed robotic system. The fixture holder in liquid filling system is designed to maintain precise dimensional

Fig. 6.3 Automated ultrasonic sealing (welding) mechanism to fill high-viscous liquid

and location tolerances between fixture holders and containers to keep reliable systematic function in automated and high-speed liquid filling process. When empty container is driven to the filling position, the indexing conveyor stops via programmable logic control (PLC) by detecting the fixture holder location signaled from installed opposed mode sensor. Horizontal air slider installed in fixture holder system moves gripper pair forward to secure container for filling process. The liquid filling nozzle moves down by vertical air slider into container entrance to begin filling process. As soon as filling process is completed, the opposed mode sensor sends the signal to PLC system and vertical air slide moves filling nozzle up. Then the gripper pairs release the container and move away from fixture holder location for next liquid filling cycle. The PLC system with opposed mode sensors in this prototype can control filling pump and nozzle to precisely determine the liquid amount being filled. When targeted liquid volume is reached, the pump and nozzle will turn off instantly resulting in precise filling process for high-viscous liquid products. The PLC system storing all operation parameters can be used for cost-effective and fast tooling changeovers. The automated and high-speed plastic welding is a fast production process to join the plastic parts together. It is one of the common processes to weld the plastic materials. Some plastic welding methodologies applied in industry include ultrasonic welding, extrusion welding, hot gas welding, high-frequency welding, injection welding, friction welding, solvent welding, speed tip welding, laser welding, contact welding, and hot plate welding. Among these plastic welding methodologies, the regular plastic welding methodology is to use external hot resource to heat joining parts together for sealing effect including extrusion welding, hot gas welding, speeding tip welding, hot plate welding, and contact welding (Song and Li 2013). The high-frequency welding technology is to join plastic components together through high-frequency electro-magnetic process including ultrasonic welding; the laser welding method is to weld work pieces together by applying pressure while the laser beam travels along welding curves; the solvent welding process applies a melt or a liquefied method to weld components together; and the friction welding method utilizes the vibration among connecting surfaces by adjusting defined vibration frequencies and amplitudes (Wang et al. 2013). The ultrasonic welding, with vibration produced from high-frequency sound energy to dissolve the plastic parts, is applied to the sealing mechanism in this new automated and high-speed production system since it is the quickest welding method suitable for high-speed production. As the ultrasonic vibration stops, molten plastics become solidified and plastic components are welded. Also the ultrasonic welding in this sealing system not only maintains high welding rate but also prevents the containers from usual damages by traditional sealing methods that relies on mechanical tolerance to control clearance between two mating parts. The sequential operation of this automated and high-speed welding system using ultrasonic welding is described as follows. As container stops at the accurate location below horn of welding mechanism, the fixture gripper travels forward to grab and secure the container for welding process. The tooling gripper picks up a container cap from vibration bowl rail and vertically moves up by pneumatic air slider. The pneumatic slider in horizontal setup drives cap gripper toward the center of container and pneumatic slider in vertical setup moves welding mechanism downward with pneumatic air rotary rotating cap gripper in 180°. As container cap is brought to the center of container top entrance by horizontal slider, vertical slider moves top tooling gripper downward to finally insert the cap into container. After cap being pressed down to the target position inside container, the top tooling gripper frees cap and travels back to pick up next cap in vibration bowl rail. The newly developed mechanism drives the ultrasonic welding horn downward and quickly welds cap and container together to seal the product.

Figures 6.4, 6.5, and 6.6 display newly developed automated and high-speed production system for chemical gas charging process based on author's new research.

The challenge of design and development in this automated and high-speed manufacturing system is to quickly and reliably seal high-pressure chamber during chemical gas charging process. The regular sealing methodologies include injecting gel for sealing purpose but show the poor sealing capability in high-speed gas charging process. In this newly developed automated gas charging and plug sealing mechanism, the plug needs to be quickly inserted to seal the container entrance while chemical gas is being charged into the container through air chamber since the pressured chemical gas will escape if container is not being sealed simultaneously. Because of this, the chemical gas charging process in this newly designed system is done inside the confined gas chamber area, shown in Fig. 6.7, and sealing plug is automatically assembled at the container inlet precisely with proper manufacturing tolerance control for best gas charging and sealing functions.

The unit chamber of sealing plug insert and gas charge in this automated and high-speed chemical gas charging system includes plug delivery inlet, central hole for plug insert and assembly, and pneumatic valves and fittings. When empty container is automatically driven to the assembly station, the linear air actuator

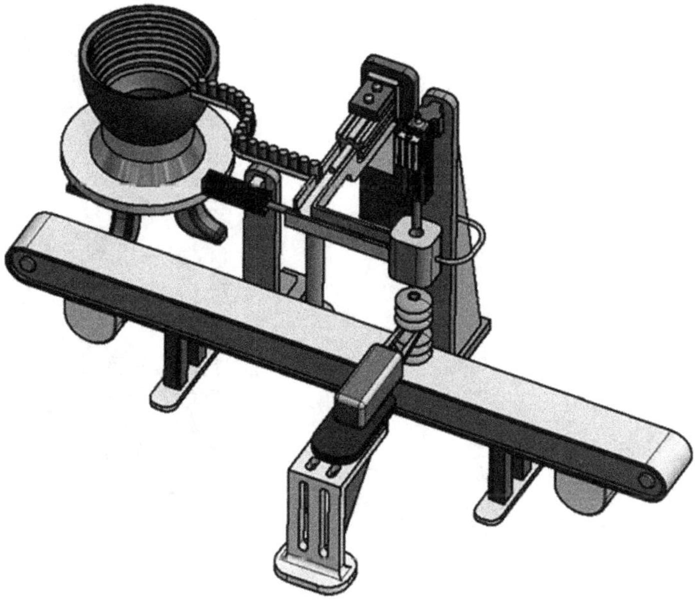

Fig. 6.4 Automated and high-speed chemical gas charging and sealing system

Fig. 6.5 Automated and high-speed sealing plug delivery mechanism

moves full plug assembly system down until it touches the inlet of container. Meanwhile, the plugs are fed to the plug assembly station by automated delivery mechanism. The sealing plug is pressed through the central hole of air chamber by top pusher with 0.125 in. below the top surface of container and pressured chemical gas is then charged into chamber. When expected gas volume is filled in the container, the top pusher continues moving plug down to fully seal the container, as shown in Fig. 6.8.

Fig. 6.6 Automated and high-speed chemical gas charging and mechanical sealing mechanism

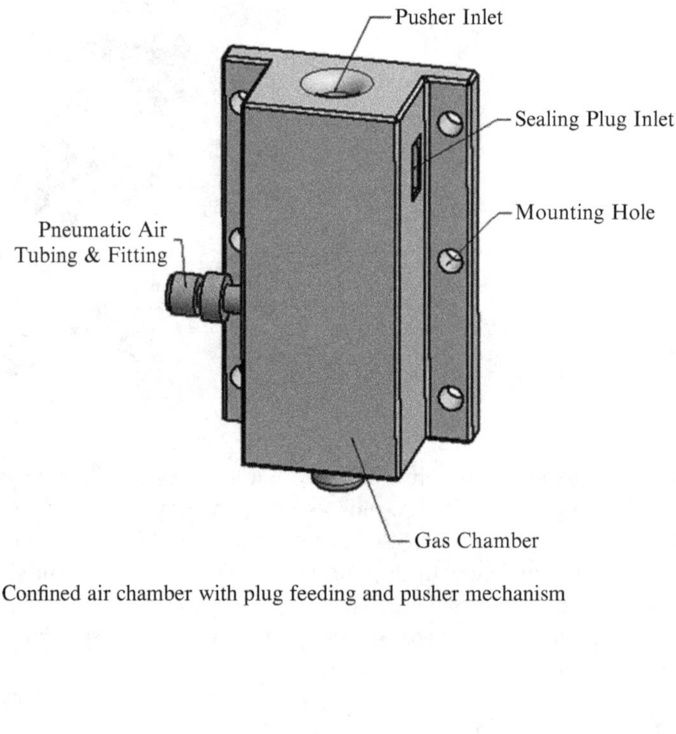

Fig. 6.7 Confined air chamber with plug feeding and pusher mechanism

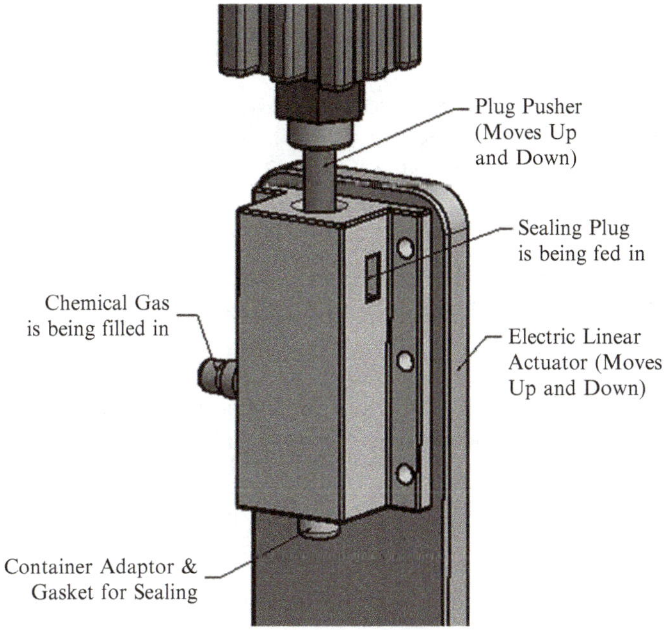

Fig. 6.8 Gas chamber and sealing plug mechanism in this new automated gas charging system

Fig. 6.9 Bronze metal ring seal

To reduce the gas leakage between entrance of container and plug insert assembly mechanism during automated and high-speed production process, a bronze metal ring, shown in Fig. 6.9, is installed to seal the leaking gap. This metal ring is tightly pressed and accurately fit to the bottom center hole in plug insert assembly mechanism.

Since the friction force between plug and center hole in plug assembly mechanism is larger than gas pressure force, there is no gas leakage from top area of plug assembly unit.

6.2 Computer-Aided Simulation of Automated and High-Speed Manufacturing Systems

The computer-aided modeling and numerical simulation of automated and high-speed high-viscous liquid filling system and automated and high-speed chemical gas charging system are described in Sects. 3.2.1 and 3.2.2, respectively.

6.2.1 Computer-Aided Simulation on Automated High-Viscous Liquid Filling System

The computational analysis and numerical simulation of high-viscous liquid leaking rate R_L in this automated and high-speed liquid filling system can be defined in the following mathematical equation (Kundu and Cohen 2008):

$$R_L = V \times E \times \left(\frac{T_P}{P_P}\right) \times {D_C}^{2.5} \times E_C \times \left(\frac{{P_{Entrance}}^2 - {P_{Exit}}^2}{L_C \times L_{SG} \times T_E \times C_F \times D_W}\right)^{0.5} \quad (6.1)$$

Here, V—liquid flow rate, E—constant, T_E—average temperature, T_P—basic temperature, D_C—container inside diameter, E_C—container filling efficiency, $P_{Entrance}$—pressure at the entrance, P_{Exit}—pressure at the exit, D_W—Darcy-Weisbach frictional coefficient, L_{SG}—specific gravity of liquids, L_C—length of container, P_P—pressure base, and C_F—coefficient of compressibility.

The mathematic equation (6.1) can be modeled in the computer-aided design and numerical simulation. It indicates that the leakage of high-viscous fluid in automated and high-speed filling process is related to different factors including liquid flow rate V and pressure difference of ($P_{Entrance} - P_{Exit}$). The computational simulation in Fig. 6.10 states that the leakage of heavy viscous liquid in this new automated and high-speed filling system is so small that the leakage can be neglected.

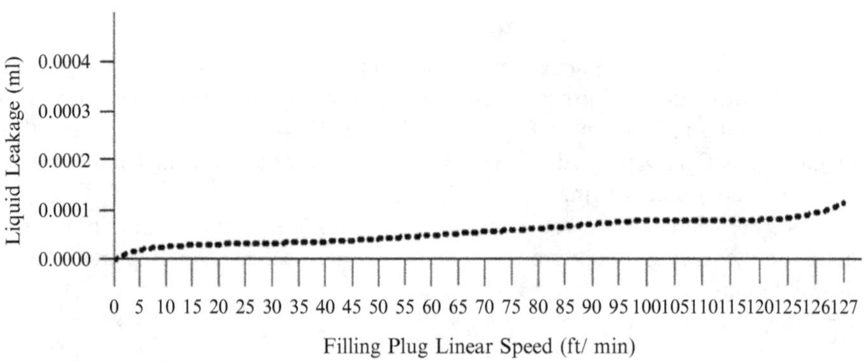

Fig. 6.10 Heavy viscous liquid leakage vs. piston linear speed

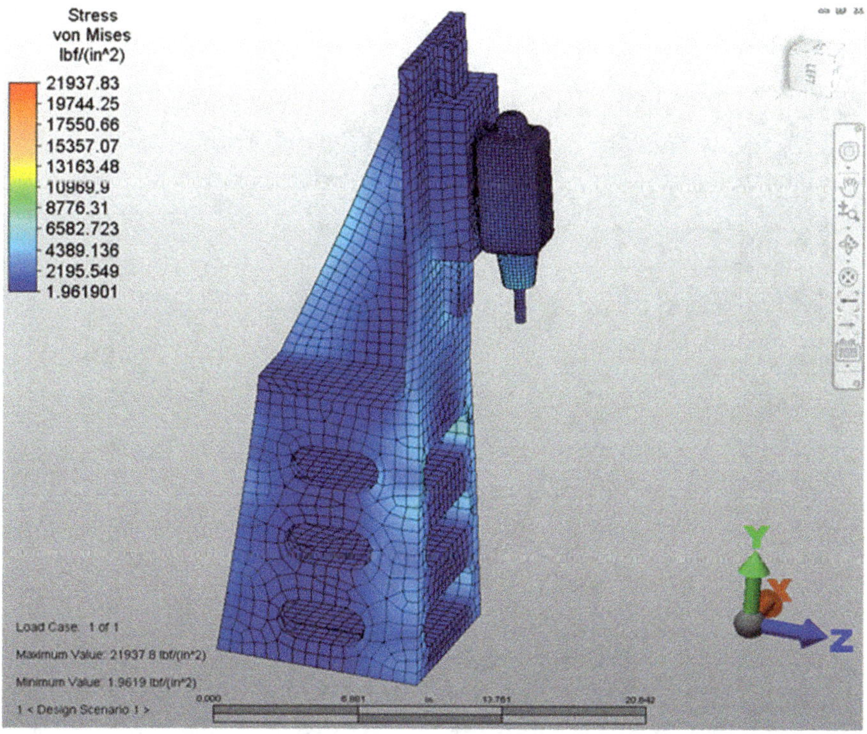

Fig. 6.11 Computational modeling of stress profile in liquid filling mechanism

The computer-aided structural and stress analysis has been performed to help this new system design and development. The analytic results on some important components, such as liquid filling mechanism and ultrasonic welding (sealing) mechanism in this new automated and high-speed viscous liquid filling system, are depicted in Figs. 6.11, 6.12, 6.13, and 6.14.

The computer-aided simulation and analysis in Figs. 6.11 and 6.12 show the stress and deflection of liquid filling mechanism in this newly designed automated and high-speed viscous liquid filling system. The analytic results present that the maximum stress of 21,937.83 psi in this liquid filling mechanism is less than the material yield strength of 36,300 psi and maximum deflection of 0.01324 in. is within material allowable deflection limit.

The computer-aided simulation and analysis in Figs. 6.13 and 6.14 indicate the stress and deflection of ultrasonic welding mechanism in this new automated and high-speed viscous liquid filling system. The analytic results present that the maximum stress of 24,973.92 psi in this ultrasonic welding mechanism is less than the material yield strength of 36,300 psi and maximum deflection of 0.01089 in. is within material allowable deflection limit.

Based on the above computer-aided modeling and numerical simulation, this new automated and high-speed high-viscous liquid filling system can function

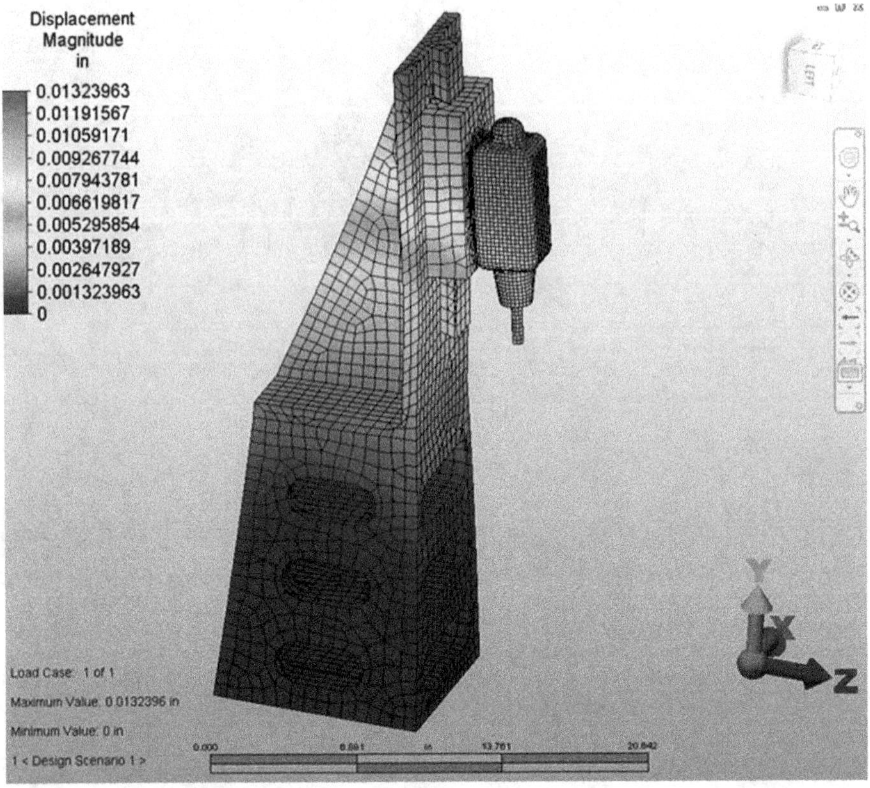

Fig. 6.12 Computational modeling of deflection profile in liquid filling mechanism

properly with good sealing capability and high production rates. Also the manufacturing cost for this new liquid filling system is relatively low since the less tolerance control is required in this new liquid filling system design.

6.2.2 Computer-Aided Simulation on Automated Chemical Gas Charging System

The mathematical modeling of gas leakage in chemical gas charging process can be expressed as follows (White 2003):

$$V_{\mathrm{GL}} = B \times \left(\Delta \frac{P}{N}\right) \times \left(\frac{1}{\Delta t}\right) \times \left(\frac{V_{\mathrm{C}}}{T}\right) \tag{6.2}$$

Here, V_{GL}—leakage rate of gas, B—unit conversion constant, $\Delta \frac{P}{N} = \left[\frac{P_{\mathrm{F}}}{N_{\mathrm{F}}} - \frac{P_{\mathrm{I}}}{N_{\mathrm{I}}}\right]$—final pressure P_{F} divided by final gas deviation constant N_{F} minus initial pressure P_{I} divided by initial gas deviation constant N_{I}, $\frac{1}{\Delta t}$—time duration in minutes to build a

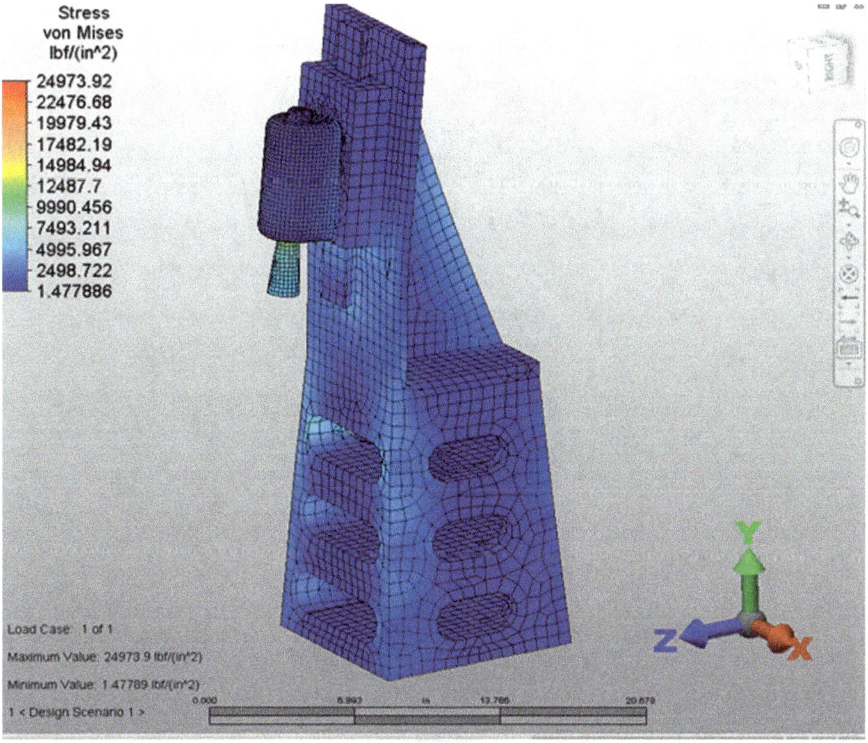

Fig. 6.13 Computational modeling of stress profile in ultrasonic welding mechanism

stabilized pressure, V_C—tubing string volume above container, and T—temperature in container.

This mathematical equation can be modeled and analyzed by computer-aided 3D modeling and numerical simulation, with analytic result expressed in Fig. 6.15.

The computational simulation result indicates that this newly designed chemical gas charging system has strong sealing capability and even very small gas leakage under high pressure can be reasonably ignored.

To perform computer-aided structural analysis, the following mathematic equation is applied to define the load or the force increment in the computational simulation (White 2003):

$$F_{\text{Load}} = P_{\text{Gas}} \times S_{\text{Area}} \tag{6.3}$$

Here, F_{Load}—load or force applied to the component, P_{Gas}—maximum gas pressure, and S_{Area}—gas charging area.

The computer-aided structural analysis can assist this new system design with analytic results on some critical parts including container fixture mechanism, plug delivery mechanism, plug insertion mechanism, and sealing plug feeding unit represented in Figs. 6.16, 6.17, 6.18, 6.19, 6.20, 6.21, 6.22, and 6.23.

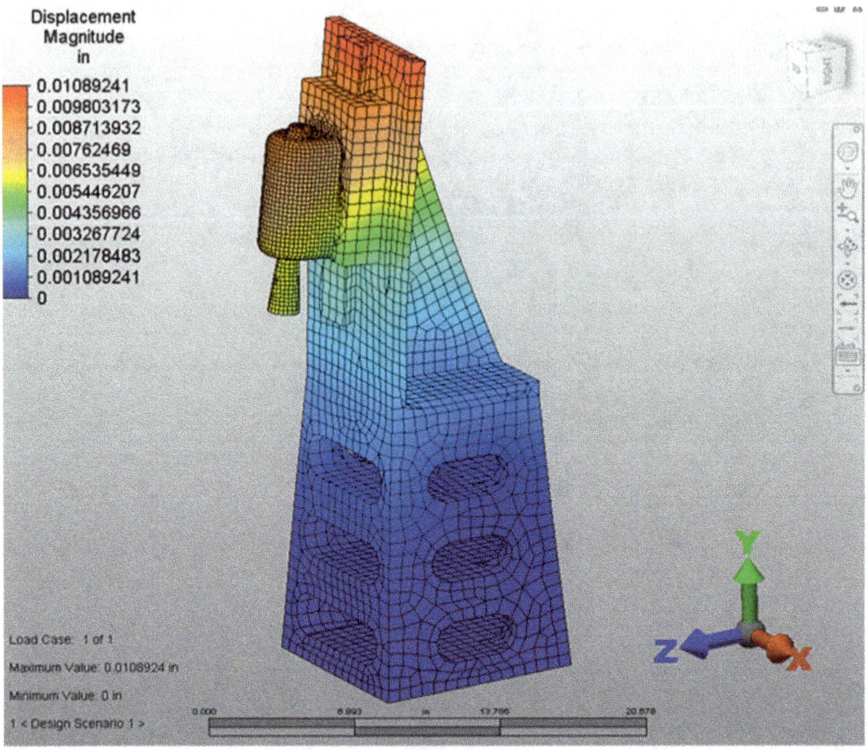

Fig. 6.14 Computational modeling of deflection profile in ultrasonic welding mechanism

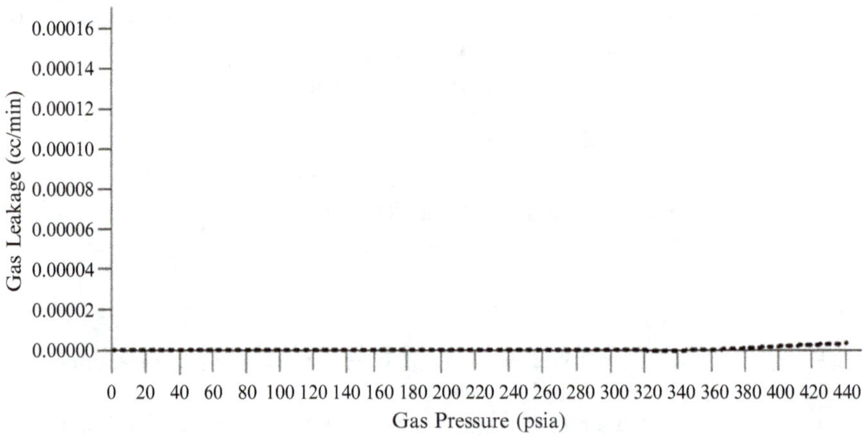

Fig. 6.15 Gas leakage vs. gas pressure

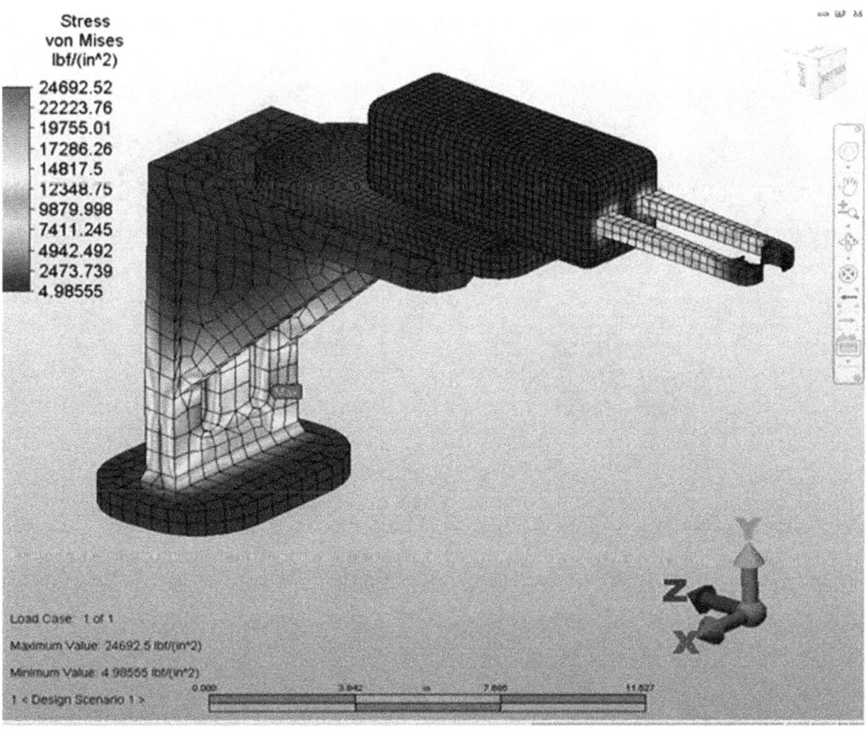

Fig. 6.16 Computational simulation of stress profile in container fixture mechanism

The computer-aided simulation and analysis in Figs. 6.16 and 6.17 display the stress and deflection of container fixture mechanism in this new automated and high-speed viscous liquid filling system. The analytic results state that the maximum stress of 24,692.52 psi in this container fixture mechanism is less than the material yield strength of 36,300 psi and maximum deflection of 0.11857 in. is within material allowable deflection limit.

The computer-aided simulation and analysis in Figs. 6.18 and 6.19 show the stress and deflection of plug delivery mechanism in this new automated and high-speed viscous liquid filling system. The analytic results demonstrate that the maximum stress of 21,072.31 psi in this plug delivery mechanism is less than the material yield strength of 36,300 psi and maximum deflection of 0.08751 in. is within material allowable deflection limit.

The computer-aided simulation and analysis in Figs. 6.20 and 6.21 indicate the stress and deflection of plug insertion mechanism in this new automated and high-speed viscous liquid filling system. The analytic results state that the maximum stress of 13,766.81 psi in this plug insertion mechanism is less than the material yield strength of 36,300 psi and maximum deflection of 0.07182 in. is within material allowable deflection limit.

The computer-aided simulation and analysis in Figs. 6.22 and 6.23 present the stress and deflection of plug feeding unit in this new automated and high-speed

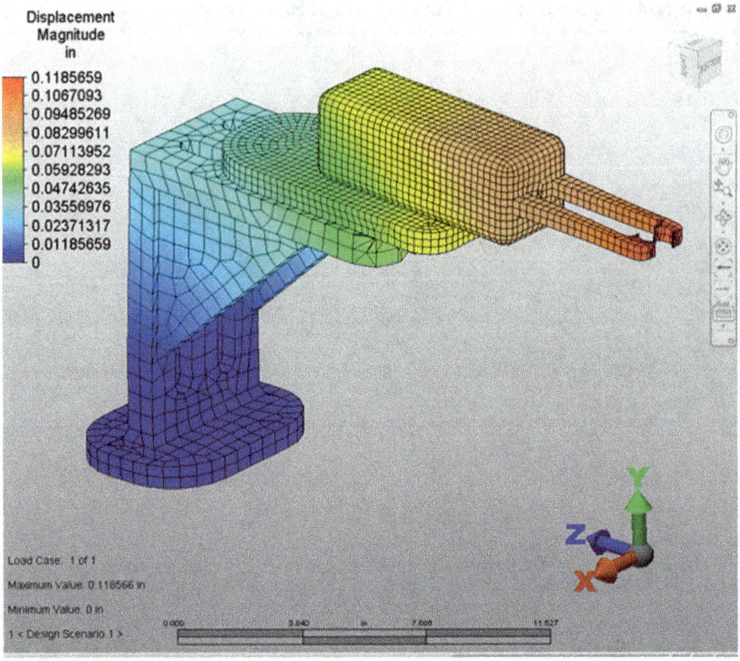

Fig. 6.17 Computational simulation of deflection profile in container fixture mechanism

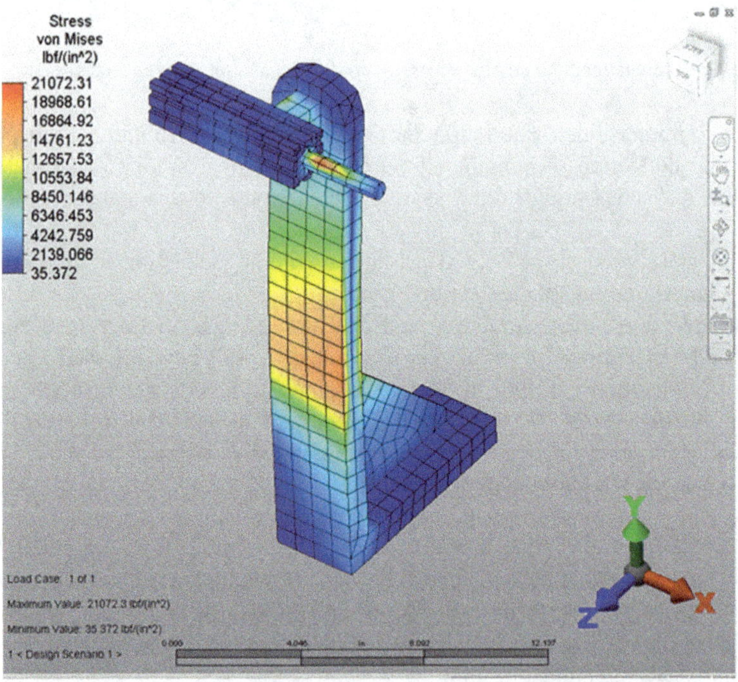

Fig. 6.18 Computational simulation of stress profile in plug delivery mechanism

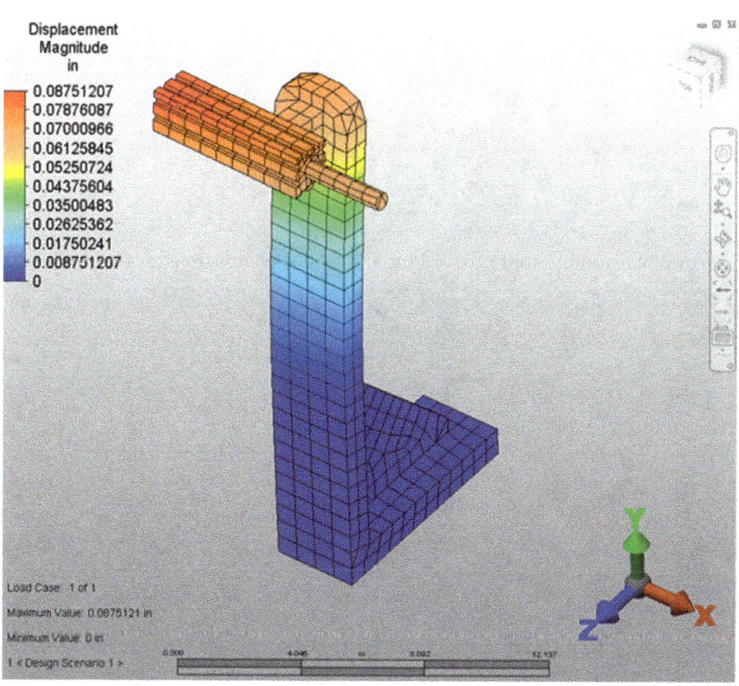

Fig. 6.19 Computational simulation of deflection profile in plug delivery mechanism

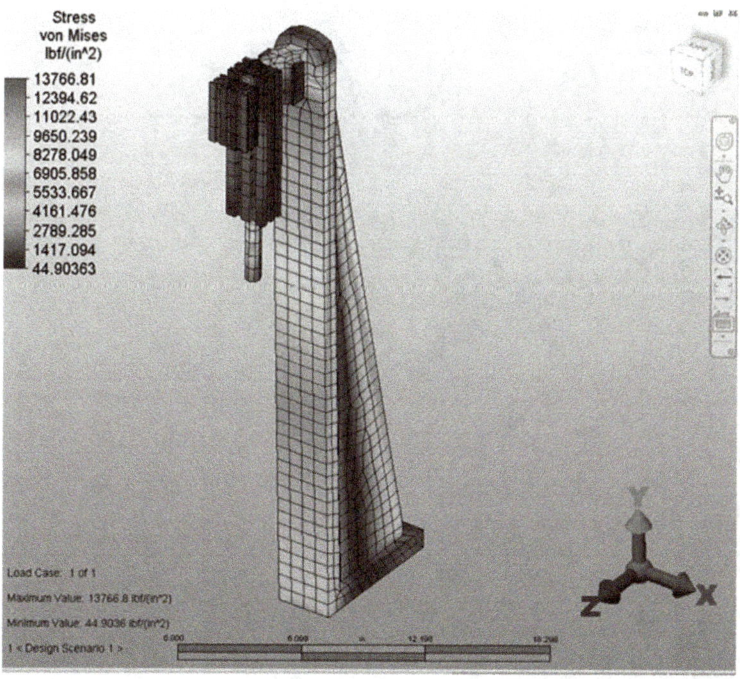

Fig. 6.20 Computational simulation of stress profile in plug insertion mechanism

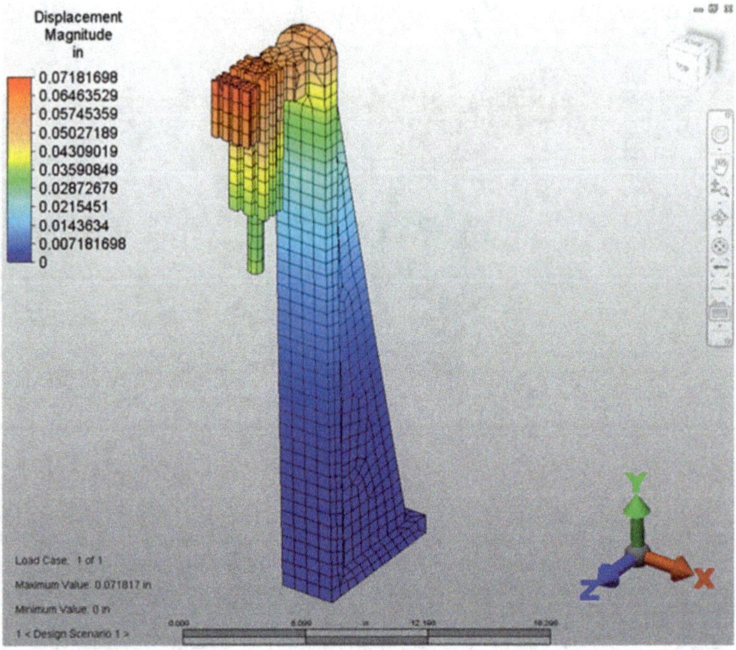

Fig. 6.21 Computational simulation of deflection profile in plug insertion mechanism

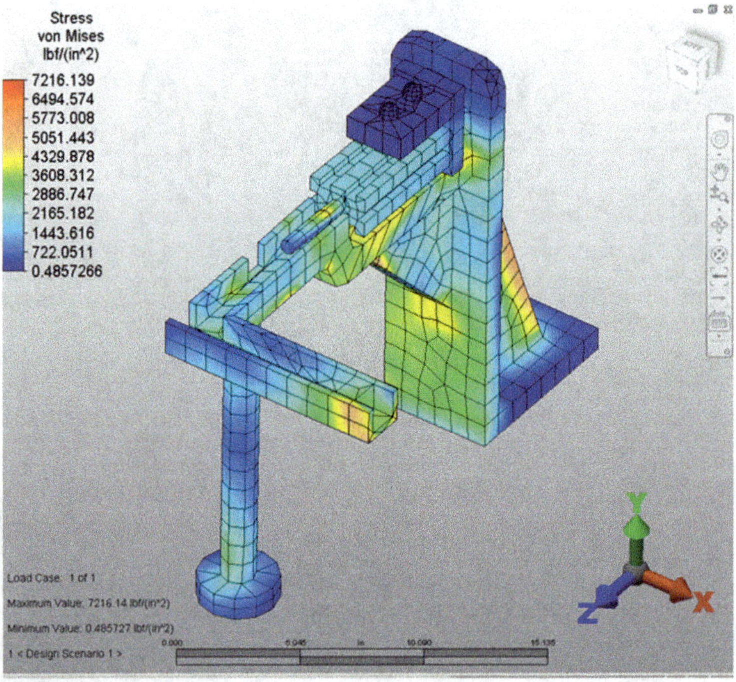

Fig. 6.22 Computational simulation of stress profile in plug feeding unit at bowl exit

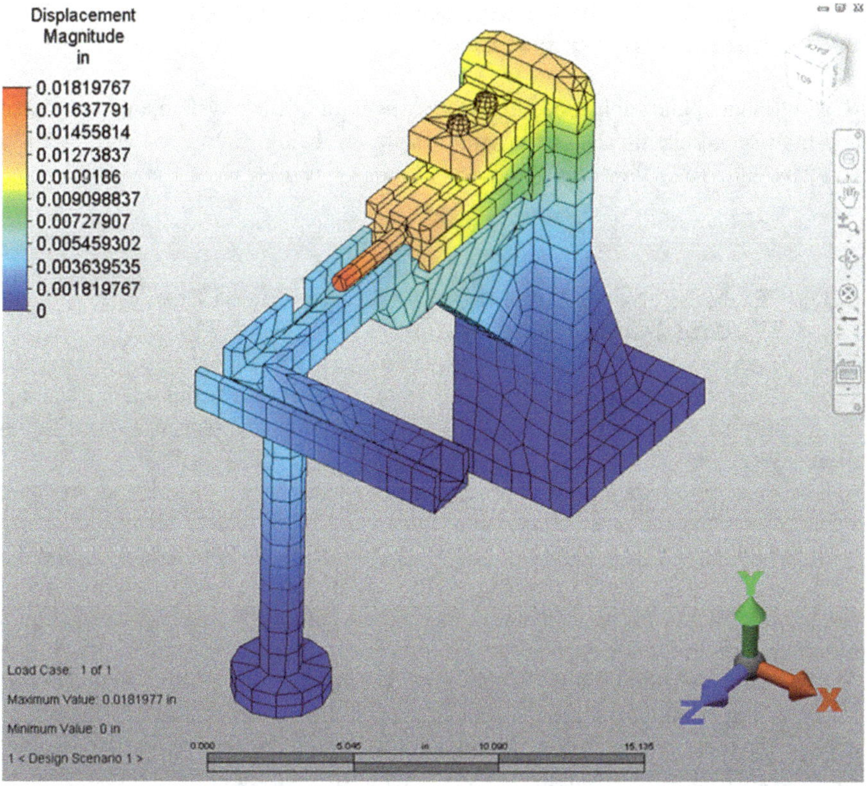

Fig. 6.23 Computational simulation of deflection profile in plug feeding unit at bowl exit

viscous liquid filling system. The analytic results state that the maximum stress of 7,216.14 psi in this plug feeding unit is less than the material yield strength of 36,300 psi and maximum deflection of 0.01819 in. is within material allowable deflection limit.

Based on computer-aided simulation in Figs. 6.16, 6.17, 6.18, 6.19, 6.20, 6.21, 6.22, and 6.23, the maximum stresses produced in all critical parts including container fixture mechanism, plug delivery mechanism, plug insertion mechanism, and sealing plug feeding unit are all less than the material yield stress and also relevant maximum deflections of these parts are within allowable limits of the materials. The above computational simulation shows the good performance of this new automated and high-speed chemical gas charging system with superior sealing capability in normal pressure and neglected gas leakage in high gas pressure. This new gas charging system can be potentially applied in high speed and fully automated manufacturing process with the cost-effective features due to simplified and less tolerance-controlled system design.

6.3 Experiment on Automated and High-Speed Manufacturing Systems

New automated and high-speed heavy viscous liquid filling system and automated and high-speed chemical gas charging system are being prototyped and tested to compare and verify the computer-aided simulation results.

6.3.1 Experiment on Automated and High-Speed Heavy Viscous Liquid Filling System

Table 6.1 records the prototype testing results of heavy viscous media leakage vs. filling plug linear speed in this new automated and high-speed viscous liquid filling system.

The prototype experiment of viscous media leakage vs. plug linear speed in Table 6.1 verifies the proper function of this liquid filling system because the close result is laid out in Fig. 6.10 by computer-aided modeling and numerical simulation.

Table 6.2 states the experiment results of maximum stress and maximum deflection of liquid filling mechanism in this new automated and high-speed liquid filling system.

The prototype testing results in Table 6.2 confirm the appropriate function of this liquid filling mechanism since the average maximum stress 21,937.78 psi and average maximum deflection 0.0128 in. are closely equal to the results of maximum stress 21,937.83 psi and maximum deflection 0.0132 in. that are recorded, respectively, in Figs. 6.11 and 6.12 by computational simulation.

Table 6.3 demonstrates the prototype experimental results of maximum stress and maximum deflection of ultrasonic welding mechanism in this new automated and high-speed liquid filling system.

Table 6.1 Prototype testing of viscous media leakage vs. filling plug linear speed

Filling plug linear speed (ft/min)	Heavy viscous media leakage (ml)
10	0.000022
20	0.000032
30	0.000038
40	0.000042
50	0.000045
60	0.000051
70	0.000068
80	0.000075
90	0.000088
100	0.000095
110	0.000102
120	0.000108

Table 6.2 Prototype testing of maximum stress and maximum deflection of liquid filling mechanism in this new heavy viscous liquid filling system

Number of experiment	Maximum stress (psi)	Maximum deflection (in.)
1	21,937.73	0.0129
2	21,937.77	0.0135
3	21,937.95	0.0122
4	21,937.91	0.0125
5	21,937.88	0.0127
6	21,937.68	0.0133
7	21,937.75	0.0131
8	21,937.78	0.0125
9	21,937.69	0.0129
10	21,937.66	0.0122
11	21,937.71	0.0133
12	21,937.81	0.0126
13	21,937.64	0.0124
14	21,937.78	0.0135
15	21,937.68	0.0127
16	21,937.88	0.0123
17	21,937.85	0.0135
18	21,937.86	0.0122
19	21,937.79	0.0129
20	21,937.88	0.0135
Average	21,937.78	0.0128

Table 6.3 Prototype experiment of maximum stress and maximum deflection of ultrasonic welding mechanism in this new heavy viscous liquid filling system

Number of experiment	Maximum stress (psi)	Maximum deflection (in.)
1	24,973.88	0.0112
2	24,973.96	0.0105
3	24,973.98	0.0118
4	24,973.94	0.0115
5	24,973.89	0.0108
6	24,973.99	0.0115
7	24,973.95	0.0113
8	24,973.88	0.0118
9	24,973.99	0.0105
10	24,973.98	0.0112
11	24,973.97	0.0116
12	24,973.96	0.0117
13	24,973.98	0.0107
14	24,973.99	0.0118
15	24,973.88	0.0115
16	24,973.98	0.0107
17	24,973.99	0.0115
18	24,973.96	0.0117
19	24,973.97	0.0108
20	24,973.98	0.0107
Average	24,973.96	0.0112

Table 6.4 Prototype experimental results of gas media leakage vs. filling gas pressure

Filling gas pressure (psia)	Gas media leakage (cc/min)
0	0.00000
20	0.00000
40	0.00000
60	0.00000
80	0.00000
100	0.00000
120	0.00000
140	0.00000
180	0.00000
200	0.00000
220	0.00000
240	0.00000
260	0.00000
280	0.00000
300	0.00000
320	0.00000
340	0.00000
360	0.00000
380	0.00000
400	0.00000
420	0.00001
440	0.00002

The experimental results of ultrasonic welding mechanism in Table 6.3 prove the normal function of this ultrasonic welding mechanism as the average maximum stress 24,973.96 psi and average maximum deflection 0.0112 in. are similar to the results of maximum stress 24,973.92 psi and maximum deflection 0.0109 in. that are presented, respectively, in Figs. 6.13 and 6.14 by computer-aided simulation.

6.3.2 Experiment on Automated and High-Speed Chemical Gas Charging System

Table 6.4 demonstrates the prototype experimental results of gas media leakage vs. filling gas pressure in this new automated and high-speed chemical gas charging system.

The prototype testing of gas media leakage vs. gas pressure in Table 6.4 verifies the proper function of this gas filling system because the equally profiled result is represented in Fig. 6.15 by computer-aided modeling and numerical simulation.

Table 6.5 represents the prototype testing results of maximum stress and maximum deflection of container fixture mechanism in this new automated and high-speed chemical gas charging system.

The prototype testing results of container fixture mechanism in Table 6.5 confirm the appropriate function of this container fixture mechanism since the average

Table 6.5 Prototype testing of maximum stress and maximum deflection of container fixture mechanism in this new automated chemical gas charging system

Number of experiment	Maximum stress (psi)	Maximum deflection (in.)
1	24,692.56	0.1178
2	24,692.48	0.1181
3	24,692.45	0.1188
4	24,692.56	0.1177
5	24,692.59	0.1179
6	24,692.46	0.1180
7	24,692.49	0.1189
8	24,692.62	0.1185
9	24,692.68	0.1178
10	24,692.57	0.1181
11	24,692.56	0.1176
12	24,692.66	0.1179
13	24,692.68	0.1182
14	24,692.62	0.1188
15	24,692.51	0.1178
16	24,692.55	0.1182
17	24,692.62	0.1185
18	24,692.64	0.1188
19	24,692.56	0.1190
20	24,692.55	0.1181
Average	24,692.57	0.1182

maximum stress 24,692.57 psi and average maximum deflection 0.1182 in. are very similar to the results of maximum stress 24,692.52 psi and maximum deflection 0.1186 in. that are indicated, respectively, in Figs. 6.16 and 6.17 by computational simulation.

Table 6.6 records the prototype experimental results of maximum stress and maximum deflection of plug delivery mechanism in this new automated and high-speed chemical gas charging system.

The prototype experimental results in Table 6.6 prove the normal function of this plug delivery mechanism as the average maximum stress 21,072.36 psi and average maximum deflection 0.0880 in. are almost equal to the results of maximum stress 21,072.31 psi and maximum deflection 0.0875 in. that are laid out, respectively, in Figs. 6.18 and 6.19 by computer-aided simulation.

Table 6.7 states the testing results of maximum stress and maximum deflection of plug insertion mechanism in this new automated and high-speed chemical gas charging system.

The prototype testing results of plug insertion mechanism in Table 6.7 verify the proper function of this plug insertion mechanism because the average maximum stress 13766.88 psi and average maximum deflection 0.0722 in. are very close to the results of maximum stress 13766.81 psi and maximum deflection 0.0718 in. that are shown, respectively, in Figs. 6.20 and 6.21 by computer-aided simulation.

Table 6.6 Prototype experiment of maximum stress and maximum deflection of plug delivery mechanism in this new automated chemical gas charging system

Number of experiment	Maximum stress (psi)	Maximum deflection (in.)
1	21,072.38	0.0882
2	21,072.35	0.0878
3	21,072.29	0.0871
4	21,072.27	0.0869
5	21,072.36	0.0882
6	21,072.39	0.0885
7	21,072.42	0.0874
8	21,072.29	0.0878
9	21,072.36	0.0888
10	21,072.38	0.0882
11	21,072.42	0.0879
12	21,072.41	0.0878
13	21,072.29	0.0888
14	21,072.39	0.0889
15	21,072.37	0.0885
16	21,072.38	0.0879
17	21,072.43	0.0873
18	21,072.27	0.0887
19	21,072.38	0.0878
20	21,072.35	0.0872
Average	21,072.36	0.0880

Table 6.7 Prototype experiment of maximum stress and maximum deflection of plug insertion mechanism in this new automated chemical gas charging system

Number of experiment	Maximum stress (psi)	Maximum deflection (in.)
1	13,766.86	0.0725
2	13,766.93	0.0716
3	13,766.83	0.0724
4	13,766.78	0.0726
5	13,766.95	0.0727
6	13,766.87	0.0716
7	13,766.77	0.0715
8	13,766.88	0.0727
9	13,766.95	0.0726
10	13,766.89	0.0724
11	13,766.92	0.0715
12	13,766.86	0.0727
13	13,766.78	0.0716
14	13,766.89	0.0725
15	13,766.86	0.0727
16	13,766.93	0.0715
17	13,766.78	0.0726
18	13,766.83	0.0713
19	13,766.88	0.0727
20	13,766.87	0.0717
Average	13,766.88	0.0722

Table 6.8 Prototype testing of maximum stress and maximum deflection of plug feeding unit in this new automated chemical gas charging system

Number of experiment	Maximum stress (psi)	Maximum deflection (in.)
1	7,216.25	0.0185
2	7,216.22	0.0175
3	7,216.18	0.0178
4	7,216.09	0.0183
5	7,216.11	0.0173
6	7,216.24	0.0175
7	7,216.27	0.0181
8	7,216.08	0.0173
9	7,216.11	0.0175
10	7,216.22	0.0173
11	7,216.24	0.0178
12	7,216.18	0.0181
13	7,216.27	0.0183
14	7,216.25	0.0172
15	7,216.22	0.0172
16	7,216.18	0.0173
17	7,216.26	0.0183
18	7,216.16	0.0173
19	7,216.18	0.0172
20	7,216.08	0.0185
Average	7,216.19	0.0177

Table 6.8 demonstrates the experimental results of maximum stress and maximum deflection of plug feeding unit in this new automated and high-speed chemical gas charging system.

The prototype experimental results of plug feeding unit in Table 6.8 confirm the appropriate function of this plug feeding unit since the average maximum stress 7,216.19 psi and average maximum deflection 0.0177 in. are closely equal to the results of maximum stress 7,216.139 psi and maximum deflection 0.0182 in. that are presented, respectively, in Figs. 6.22 and 6.23 by computer-aided simulation.

6.4 Discussion and Future Improvement on Automated and High-Speed Manufacturing Systems

Automation can bring explosive growth of business. The real-time systems will effectively control the complex adaptive and multiprocessing systems. The applied control software permits elemental nodes to communicate with each other in the larger and complicated adaptive system. Since the automated and high-speed production systems are too expensive to be rebuilt for each design change and system modification, the designed automated systems should be potentially flexible or easily reconfigurable to produce a variety of products on customized demand.

Automated and high-speed production technologies in modern workplace require skilled individuals to smoothly and correctly handle the technical problems that vary depending on the product being manufactured, process being employed, and machineries being utilized. The automated production systems needs task forces having basic production knowledge and hands-on experience to perform troubleshooting of high-tech and high-speed automated machinery systems or reinstallation of it if required to maintain continuous production flow. Because of these above technical reasons, the automated and high-speed systems are highly recommended to be designed for reliable function, quicker maintenance, simple installation, cheaper reconfiguration, and easier technical training. Two new automated and high-speed production systems of high-viscous liquid filling and chemical gas charging introduced in this book have aimed at these professional design targets. Upon a few small changes, viscous liquid filling system can be easily reconfigured for chemical gas charging process. This new cost-effective system design allows efficient quality diagnostic software and adaptable interfaces working in high-level reliability with permeate access up to bottom diagnosis and error-rectification database through centralized control. To further improve these new automated and high-speed production systems, the modifications will include making moving mechanisms (such as ultrasonic welder support, liquid filling nozzle connector, and sealing plug plunger) lighter to reduce the vibration, modifying design layout for easier maintenance, searching cheaper high-viscous liquid pump to reduce the cost, and readjusting the system tolerance control for cost-effective manufacturing.

Robotic System

7

7.1 Design of Robotic System for Industrial Applications

Robotics research integrates multiple academic and technical fields including mathematics, mechatronics, physics, automation, and mechanical engineering. Robotic systems can be applied to the industry to perform many special, dangerous, heavy labor involved, accurate, and repetitive tasks. Figure 7.1 shows the prototype of this newly designed robotic system. It can be used for automated manufacturing and assembly processes including picking up and placing the work pieces and products in automated industrial production.

Actuators in robotic system operate as human muscles to change the energy into body motion. Electric motor is a very common type of actuator, linear actuator is used when strong force is needed in linear motion, piezo motor relies on piezo element vibration to operate the motor, air wire is a hollow cylindrical element and its length varies as different internal air pressure is applied, and electroactive polymer is a plastic material and its length can be altered by electricity. The sensor has a firm central element surrounded by conductive liquids that is boxed by the elastic substance. Electrode, mounted at the hard central element surface, is linked to a resistance-inspecting unit inside the central element. As elastic substance touches an object, the liquid volume boxed electrode will vary causing the resistance change that traces the external forces from target object. Computer-aided vision technology can be applied in robotic system and automation to obtain numerical information from digital images via video and camera systems. Robotic systems can flexibly manipulate objects by robotic arms to perform different complex tasks. Mechanical grippers are commonly used in robotic system to pick up and place small objects and vacuum grippers are widely applied to pick and place large objects. Kinematics can be applied to determine effectors' moving velocity, travel acceleration, reaching location, and oriented rotation while designing and analyzing the robotic system. Dynamics can be used to determine the external force or load effect on robotic arm movement and dynamic methodology

J. Zheng Li, *CAD, 3D Modeling, Engineering Analysis, and Prototype Experimentation*, DOI 10.1007/978-3-319-05921-1_7

Fig. 7.1 Prototype of new robotic system

is an effective tool used in computational simulation to adjust and optimize the control algorithms of robotic systems.

The prototype shown in Fig. 7.1 has been analyzed and validated through computer-aided 3D modeling, numerical simulation, and prototype experiment in the following chapters. The analytic and experimental results show that the maximum 3D moving distances of this new robotic system are 8 ft (x direction), 12 ft (y direction), and 10 ft (z direction), respectively.

7.2 Computer-Aided Simulation on Robotic System for Industrial Applications

One new robotic system from author's current researches has been introduced and analyzed in this chapter for picking and placing products in biomedical/surgical procedures and automated production processes. The robotic systems is 3D modeled, designed, and validated through computer-aided modeling, numerical simulation, and engineering analysis. The designed robotic system is programmed with specific mathematical model that accurately controls the robot arm movement. The simplest double-link robot arm with less complex mathematical model is shown in Fig. 7.2.

In simplified double-lever arm system, the first lever A_B moves around the origin point O in the M_N Cartesian coordinate system and the second lever B_C goes around the hinge point B that joins levers A_B and B_C. The angle ζ is selected between lever A_B and horizontal axis O_M and angle λ is set between

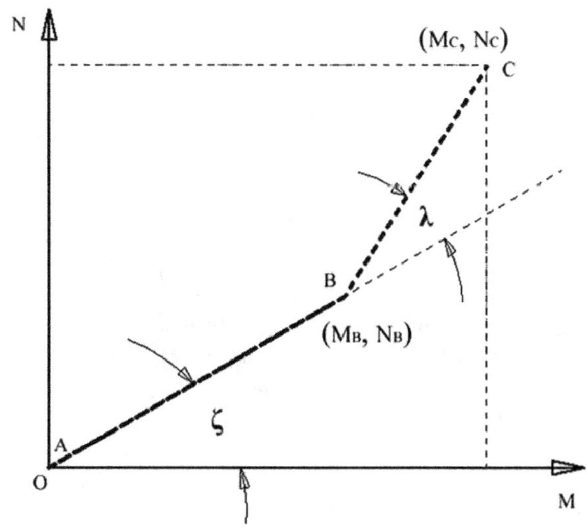

Fig. 7.2 Double-lever model for robot arm

levers A_B and B_C. The robot arm is at the end places (M_B, N_B) of lever A_B and (M_C, N_C) of lever B_C.

Three analytic methodologies can be applied to analyze the robotic system: kinematics—specify robot arm position (M_B, N_B) while angles ζ and λ are known; inverse kinematics—solve angles ζ and λ while robot arm position (M_B, N_B) is given; and trajectory generation—ascertain robot arm new position (M_C, N_C) by modifying angles ζ and λ while present position (M_B, N_B) is provided.

- Kinematical method:

From Fig. 7.2, the end point of lever B_C can be defined as follows (Gevorkian 2007):

$$M_C = (A\_B) \times \cos(\zeta) + (B\_C) \times \cos(\zeta + \lambda) \tag{7.1}$$

$$N_C = (A\_B) \times \sin(\zeta) + (B\_C) \times \sin(\zeta + \lambda) \tag{7.2}$$

Here,

$$M_B = (A\_B) \times \cos(\zeta)$$

$$N_B = (A\_B) \times \sin(\zeta)$$

$$M_C = M_B + (B\_C) \times \cos(\zeta + \lambda)$$

$$N_C = N_B + (B\_C) \times \sin(\zeta + \lambda)$$

- Inverse kinematical method:

Based on kinematical equations (7.1) and (7.2), it needs a nonlinear solution to specify angles ζ and λ while end position (M_C, N_C) is provided.

Considering the following mathematical equations (Gevorkian 2007):

$$\cos(\alpha + \beta) = \cos(\alpha)\cos(\beta) - \sin(\alpha)\sin(\beta) \tag{7.3}$$

$$\sin(\alpha + \beta) = \cos(\alpha)\sin(\beta) + \sin(\alpha)\cos(\beta) \tag{7.4}$$

the algebraic model can be used to resolve this nonlinear problem by combining mathematical equations (7.1) and (7.2). Solving the above nonlinear mathematical equations can obtain new mathematical equation to determine angle λ (Gevorkian 2007):

$$\lambda = \arccos\left[\frac{M_C{}^2 + N_C{}^2 - (\text{A\_B})^2 - (\text{B\_C})^2}{2 \times (\text{A\_B}) \times (\text{B\_C})}\right] \tag{7.5}$$

Since $\cos(\lambda) = \cos(-\lambda)$, the inverse kinematical problem can have two solutions with lever B_C clockwise and counterclockwise rotation as displayed in Fig. 7.3.

Angles ζ can be specified upon solution of angle λ in inverse kinematical model (Gevorkian 2007):

$$M_C = [(\text{A\_B}) + (\text{B\_C}) \times \cos(\lambda)] \times \cos(\zeta) - (\text{B\_C}) \times \sin(\lambda) \times \sin(\zeta) \tag{7.6}$$

$$N_C = (\text{B\_C}) \times \sin(\lambda) \times \cos(\zeta) + [(\text{A\_B}) + (\text{B\_C}) \times \cos(\lambda)] \times \sin(\zeta) \tag{7.7}$$

The angle λ can be specified by resolving the above nonlinear mathematical equations (7.6) and (7.7).

- Trajectory-generated method:

In double-levered robot arm application, the arm position (M_B, N_B) is correlated to the lever angles ζ_B and λ_B. As robot arm revolves to a new position (M_C, N_C), the angles ζ_C and λ_C can be specified upon inverse kinematical analysis. Several methods to manipulate robotic arm rotating from one place to another place are alter angle ζ prior to setting λ, modify angle λ before settling ζ, change angles ζ and λ together at the same time, move robot arm around in clockwise direction, and turn robot arm around in counterclockwise direction. An efficient and well-functioned robot system design can cut down the energy necessitated to move robot arm and minimize the time required to switch the systematic mechanism. Figure 7.4 lays out a robot arm orientating in a trajectory-generated curve with angle ζ altering in clockwise direction.

Two traveling trajectories of robot arm in Fig. 7.4 convey the desirable moving situation. In linear trajectory-generated method, the straight or non-straight line can be separated into multiple tiny elements and associated (M_i, N_i) coordinates at the

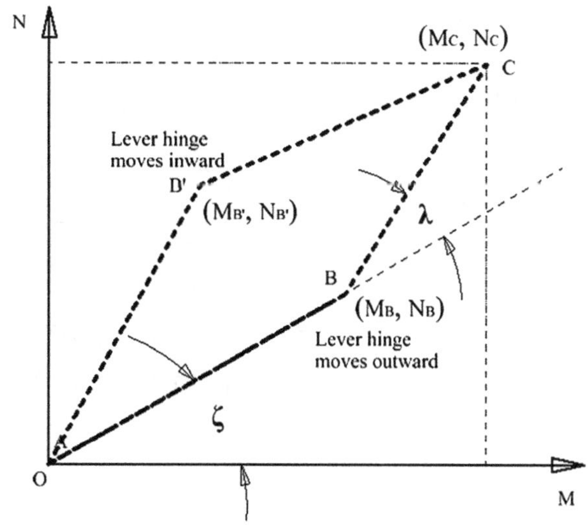

Fig. 7.3 Inverse kinematical model for robotic arm

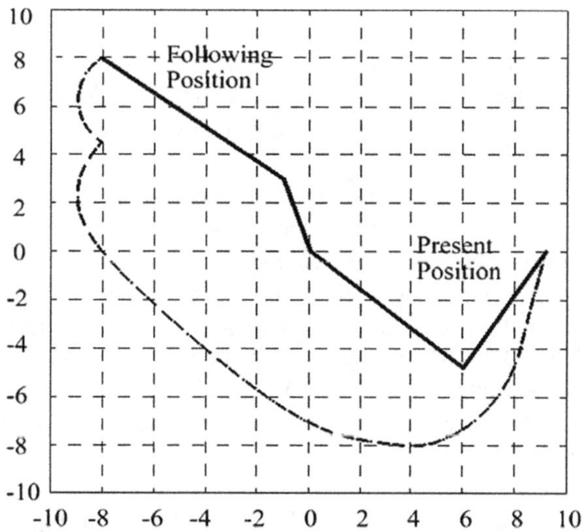

Fig. 7.4 Robot arm in trajectory-generated move with clockwise change in angle ζ

end of each element are determined via computer-aided simulation. The targeted angle pairs ζ_i and λ_i are calculated for each (M_i, N_i) coordinate pair and robot arm mechanisms can be manipulated as each new pair of ζ_i and λ_i is computed. The computational simulation of angle pair's ζ_i and λ_i is continuously performed along straight or non-straight line until anticipated arm position is achieved.

The more complex robotic arm movement can also be simulated and analyzed through computational modeling and numerical simulation. Figure 7.5 indicates the simulated trajectory moving curve of robot arm upon many important data points along traveling path. The robot arm moving path becomes smoother when angle pairs ζ_i and λ_i are being adjusted at the same time.

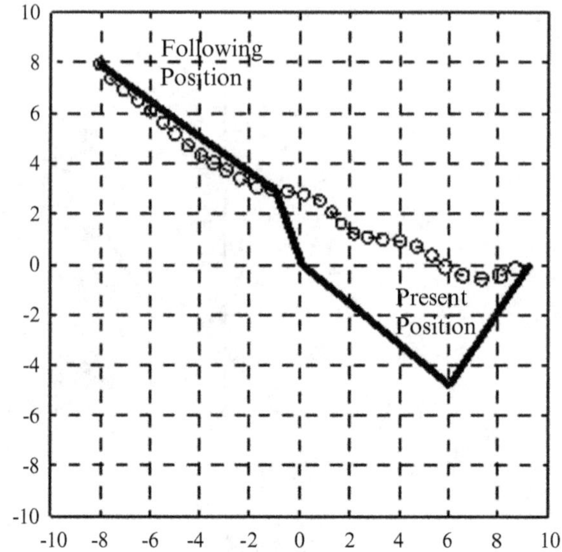

Fig. 7.5 Computational simulation of robotic arm traveling path upon multiple data points

Figure 7.6 records the robot arm traveling in an alternative trajectory path with smoother curve as angle pair's ζ_i and λ_i are manipulated simultaneously.

The robotic system is a nonlinear functioning system that demands the integration of complicated mathematical modeling, computer-aided analysis, and numerical simulation to study and resolve the robotic motion problems. The different lever lengths can be applied to specify possible 3D traveling ranges that robotic arms can cover. The inverse kinematical method can be revised to follow through the trajectory-generated solution and apply computational modeling and numerical simulation to analyze the complex motion of robotic arm system. Figures 7.7, 7.8, 7.9, 7.10, 7.11, and 7.12 show computer-aided simulation and solution of pulse-width modulation (PWM) values that are used to drive the motors to move robotic system in different directions, to determine the kinematic motion of the robotic system.

The traveling trajectories of robotic system, converted from computer-aided analytic solutions of PWM values exhibited in Figs. 7.7, 7.8, 7.9, 7.10, 7.11, and 7.12, can be used to optimize the robotic system design to achieve desired moving ranges through computer-aided modeling and numerical simulation. For each (M_i, N_i) coordinate pair, the required angles ζ_i and λ_i are calculated via computational modeling and numerical simulation for optimal systematic function. The robotic arm mechanisms are being updated as soon as each new angle pair of ζ_i and λ_i is verified and computational solutions of angles ζ_i and λ_i continue until the desired robotic arm location is reached. The computational simulation shows that the maximum 3D traveling ranges of this new robotic system shown in Fig. 7.13 are 8 ft (front to back), 12 ft (left to right), and 10 ft (top to bottom).

To verify if this newly developed robotic system has necessary strong structure to handle automated and high-speed manufacturing and production, the

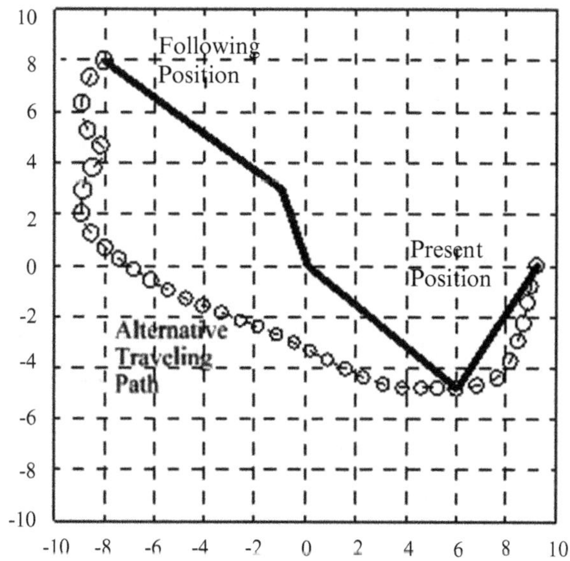

Fig. 7.6 Computational simulation of robotic arm traveling with an alternative moving path

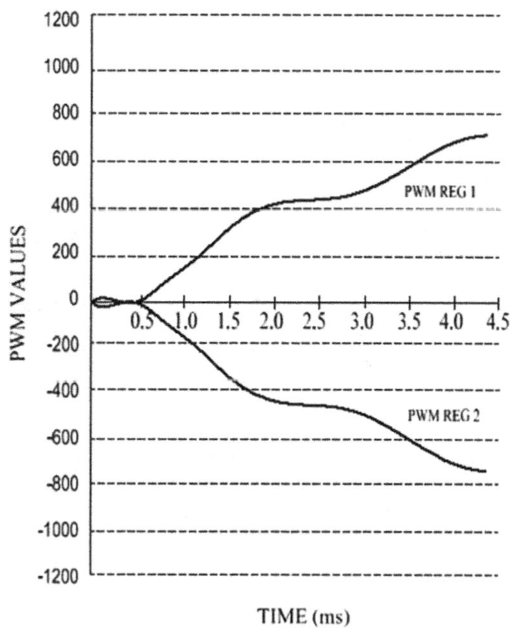

Fig. 7.7 Simulated PWM values—drive motor to move robot arm toward the front

computer-aided structural analysis has been performed on some critical components, with stress and deflection profiles presented in Figs. 7.14, 7.15, 7.16, 7.17, 7.18, and 7.19.

The computer-aided simulation and analysis in Figs. 7.14 and 7.15 show the stress and deflection of arm in this new robotic system. The analytic results tell that

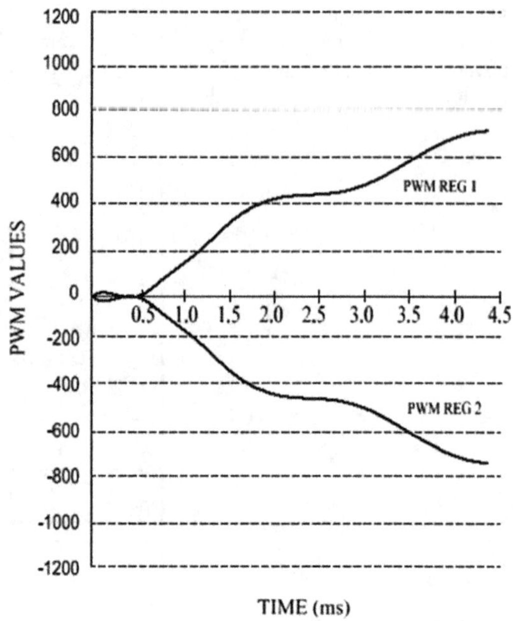

Fig. 7.8 Simulated PWM values—drive motor to move robot arm toward the back

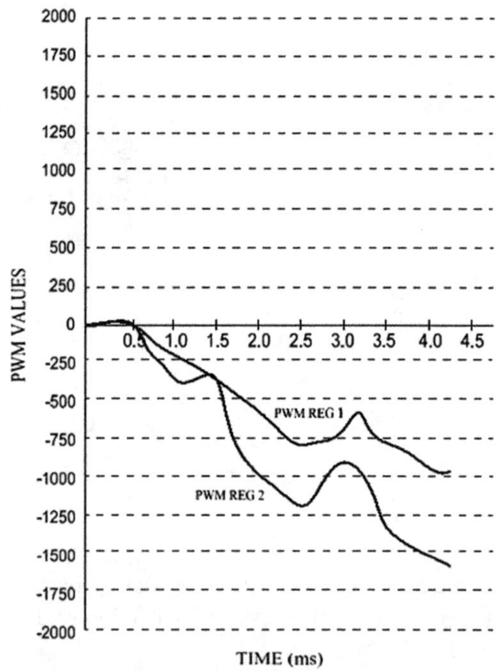

Fig. 7.9 Simulated PWM values—drive motor to move robot arm toward the left

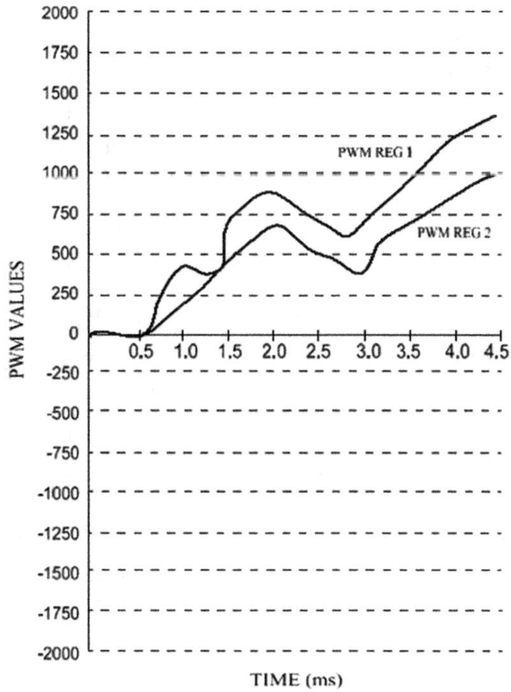

Fig. 7.10 Simulated PWM values—drive motor to move robot arm toward the right

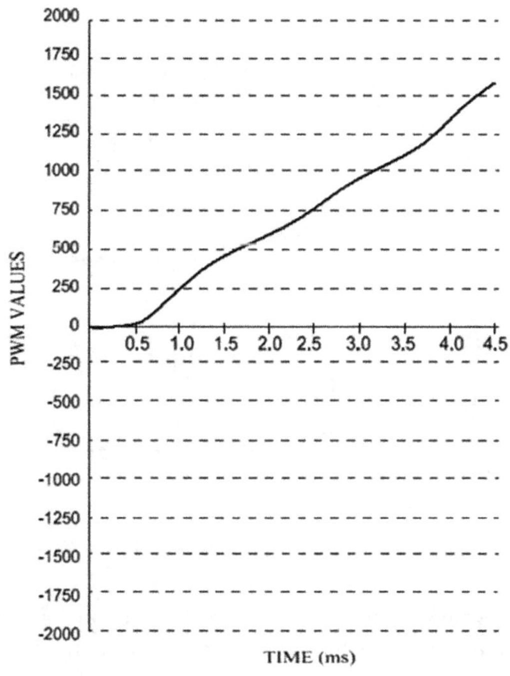

Fig. 7.11 Simulated PWM values—drive motor to move robot arm toward the upper direction

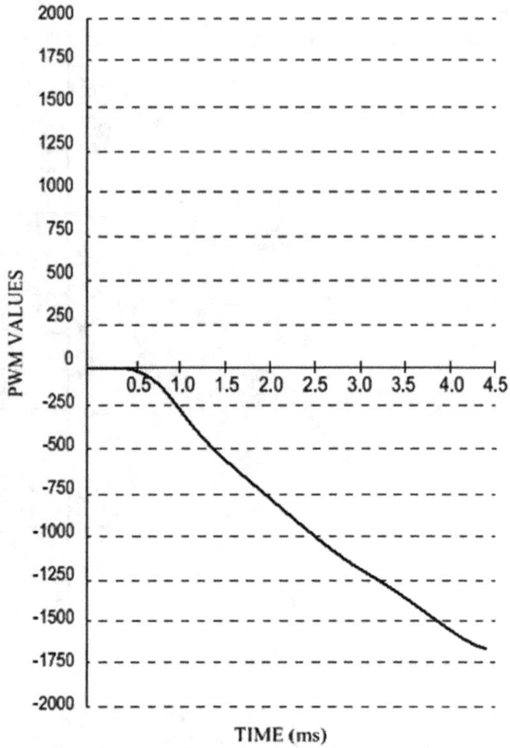

Fig. 7.12 Simulated PWM values—drive motor to move robot arm toward the lower direction

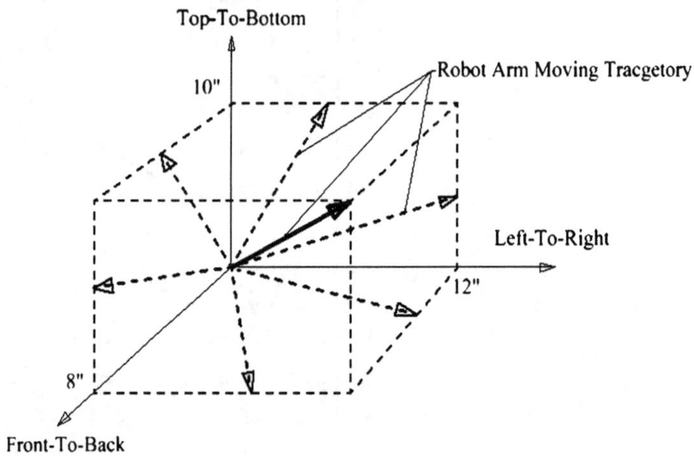

Fig. 7.13 Simulated maximum 3D moving range of newly developed robotic system

the maximum stress of 22,421.51 psi in this arm is less than the material yield strength of 36,300 psi and maximum deflection of 0.01045 in. is within material allowable deflection limit.

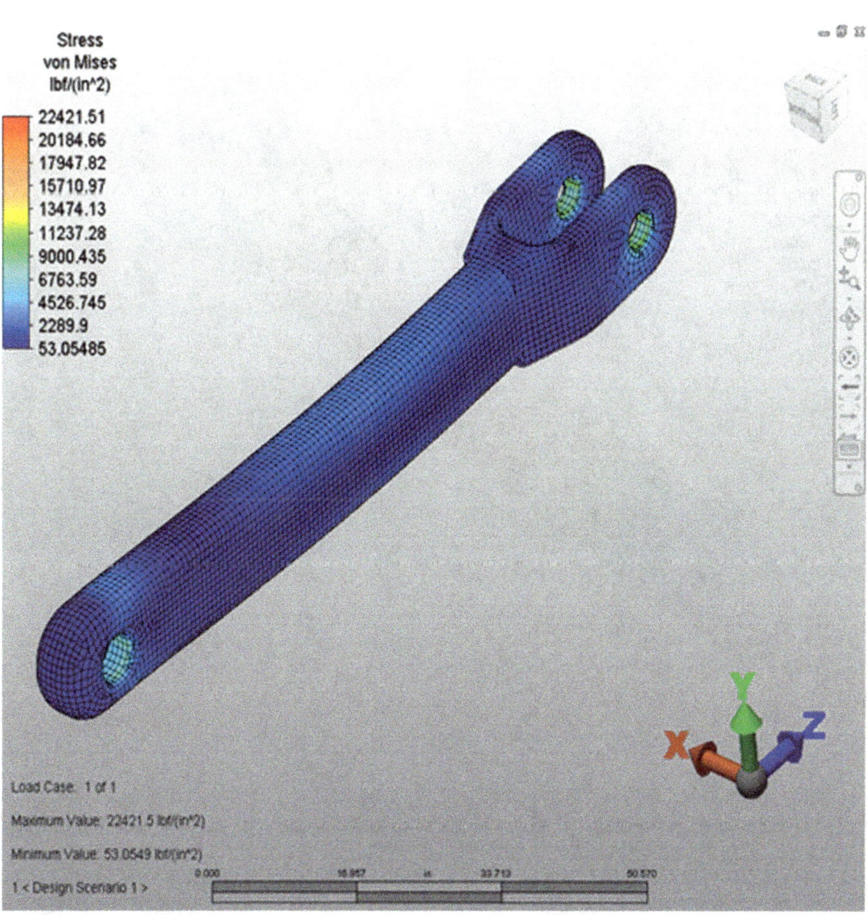

Fig. 7.14 Computer-aided simulation of stress profile in the arm of new robotic system

The computer-aided simulation and analysis in Figs. 7.16 and 7.17 present the stress and deflection of arm link in this new robotic system. The analytic results demonstrate that the maximum stress of 22,634.97 psi in this arm link is less than the material yield strength of 36,300 psi and maximum deflection of 0.04469 in. is within material allowable deflection limit.

The computer-aided simulation and analysis in Figs. 7.18 and 7.19 indicate the stress and deflection of arm base in this new robotic system. The analytic results state that the maximum stress of 19,097.23 psi in this arm base is less than the material yield strength of 36,300 psi and maximum deflection of 0.23198 in. is within material allowable deflection limit.

The above computer-aided simulation results indicated that the maximum stresses on these critical components are all below the material yield stress and maximum material deflections are all within material allowable deformation limits. The above analytic solutions have shown the proper systematic function and reliable quality of this newly designed and developed robotic system.

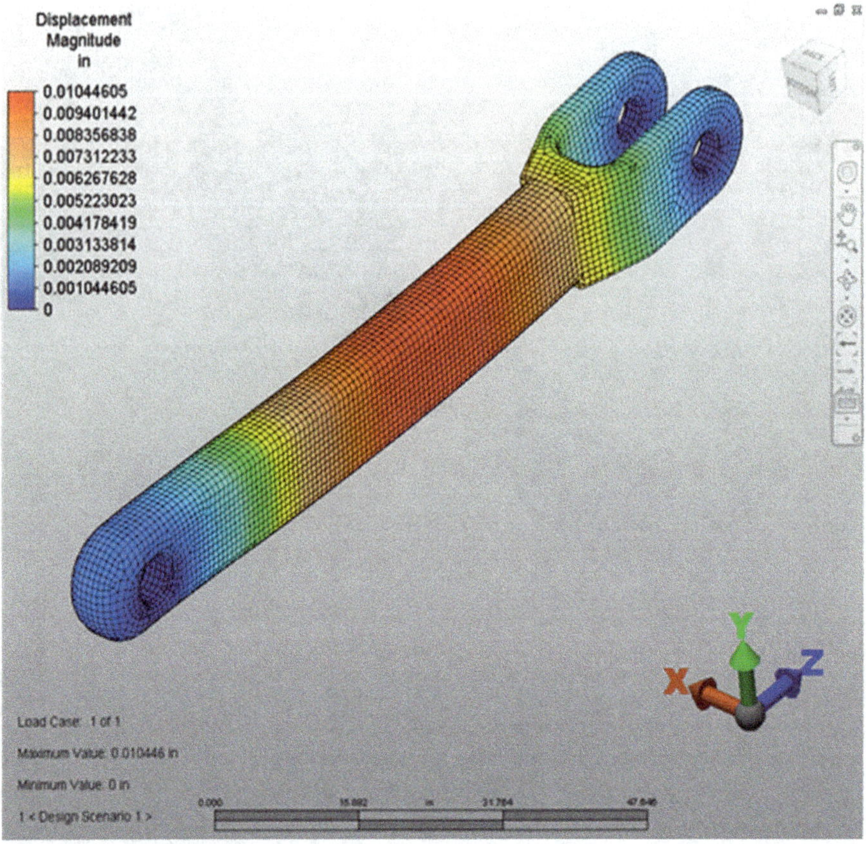

Fig. 7.15 Computer-aided simulation of deflection profile in the arm of new robotic system

7.3 Experiment on Robotic System for Industrial Applications

The kinematic motion of robot arm in this robotic system is measured by pulse-width modulation (PWM) methodology to the prototype. The PWM register's values are utilized to operate the motors of robotic system in different orientations. The following tables present the movement of the robot arm in this new robotic system toward the front, back, left, right, higher, and lower directions.

Table 7.1 demonstrates the prototype testing results of PWM register as robot arm of this new robotic system moves to the front.

Based on Table 7.1, while the robot arm of this new robotic system moves to the front, the prototype testing results of PWM register verify the proper function because the testing values are close to the profiled PWM register results by computer-aided modeling and simulation as shown in Fig. 7.7.

Table 7.2 depicts the prototype experimental results of PWM register as robot arm of this new robotic system moves to the back.

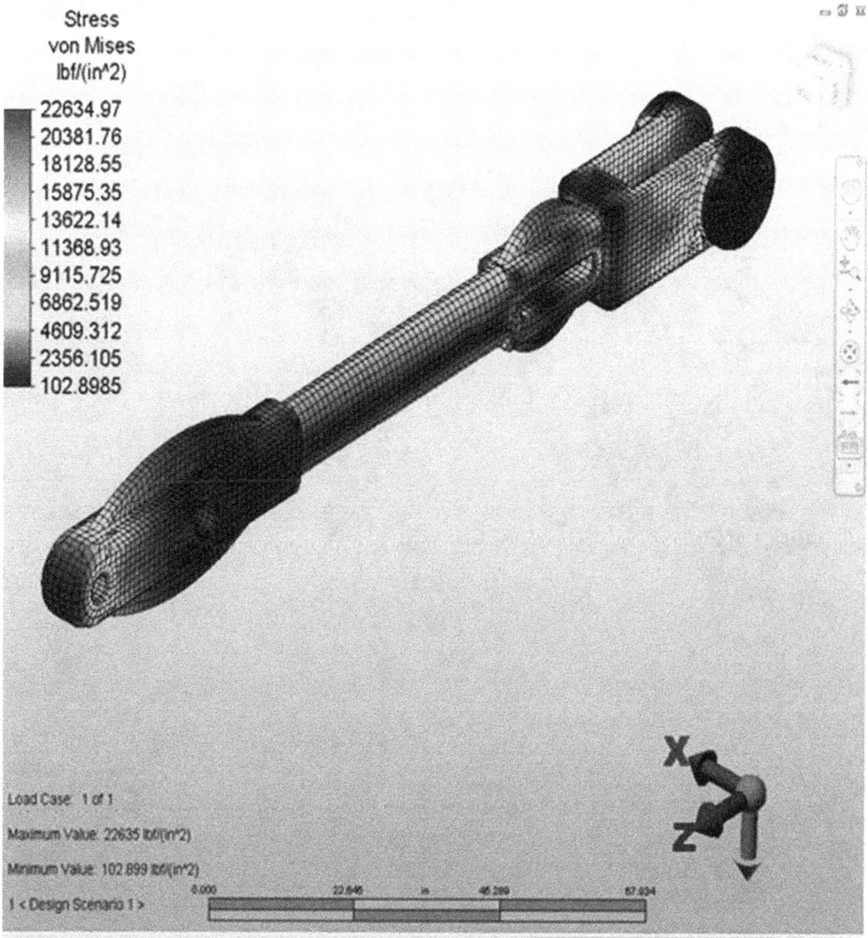

Fig. 7.16 Computer-aided simulation of stress profile in the link of new robotic system

Table 7.2 represents that while the robot arm of this new robotic system moves to the back, the prototype experimental results of PWM register confirm the appropriate function since the testing values are similar to the simulated PWM register results by computer-aided modeling and simulation as depicted in Fig. 7.8.

Table 7.3 expresses the prototype testing results of PWM register as robot arm of this new robotic system moves to the left.

Table 7.3 conveys that while the robot arm of this new robotic system moves to the left, the prototype testing results of PWM register prove the normal function as the testing values are almost equal to the simulated PWM register results by computer-aided modeling and simulation conveyed as shown in Fig. 7.9.

Table 7.4 indicates the prototype experimental results of PWM register as robot arm of this new robotic system moves to the right.

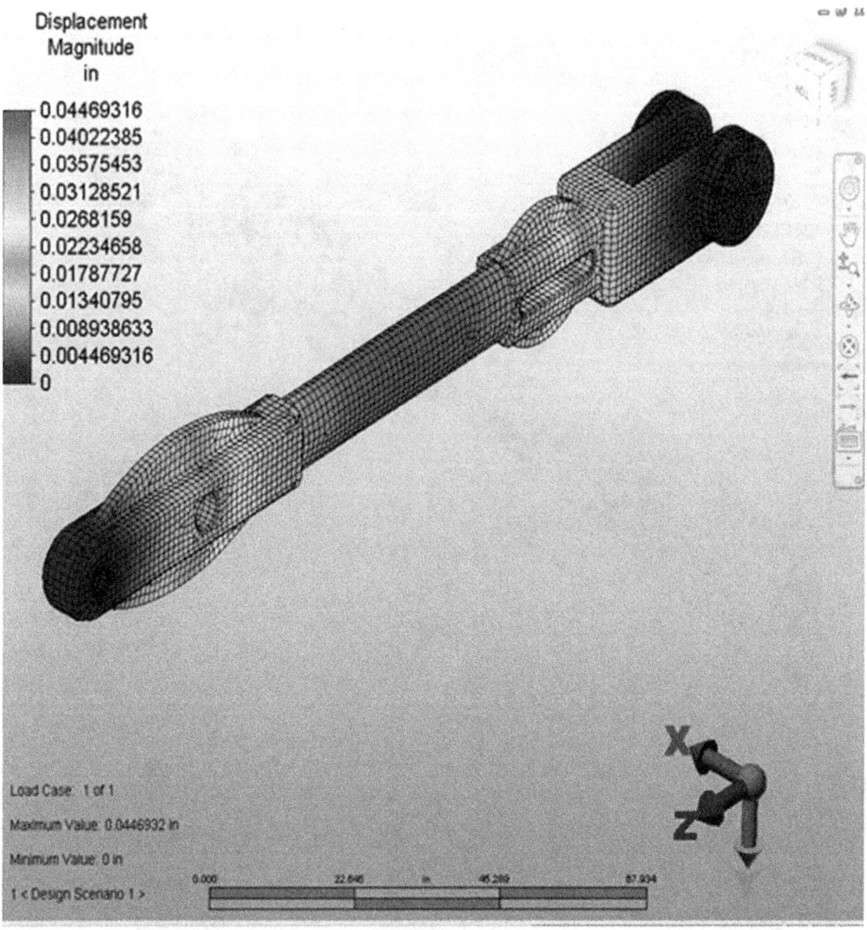

Fig. 7.17 Computer-aided simulation of deflection profile in the link of new robotic system

Table 7.4 records that while the robot arm of this new robotic system moves to the right, the prototype experimental results of PWM register verify the proper function because the testing values are very similar to the simulated PWM register results by computer-aided modeling and simulation as shown in Fig. 7.10.

Table 7.5 lays out the prototype testing results of PWM register as robot arm of this new robotic system moves to the upper direction.

Table 7.5 states that while the robot arm of this new robotic system moves to the upper direction, the prototype experimental results of PWM register confirm the appropriate function since the testing values are closely equal to the simulated PWM register results by computer-aided modeling and simulation as shown in Fig. 7.11.

Table 7.6 displays the prototype experimental results of PWM register as robot arm of this new robotic system moves to the lower direction.

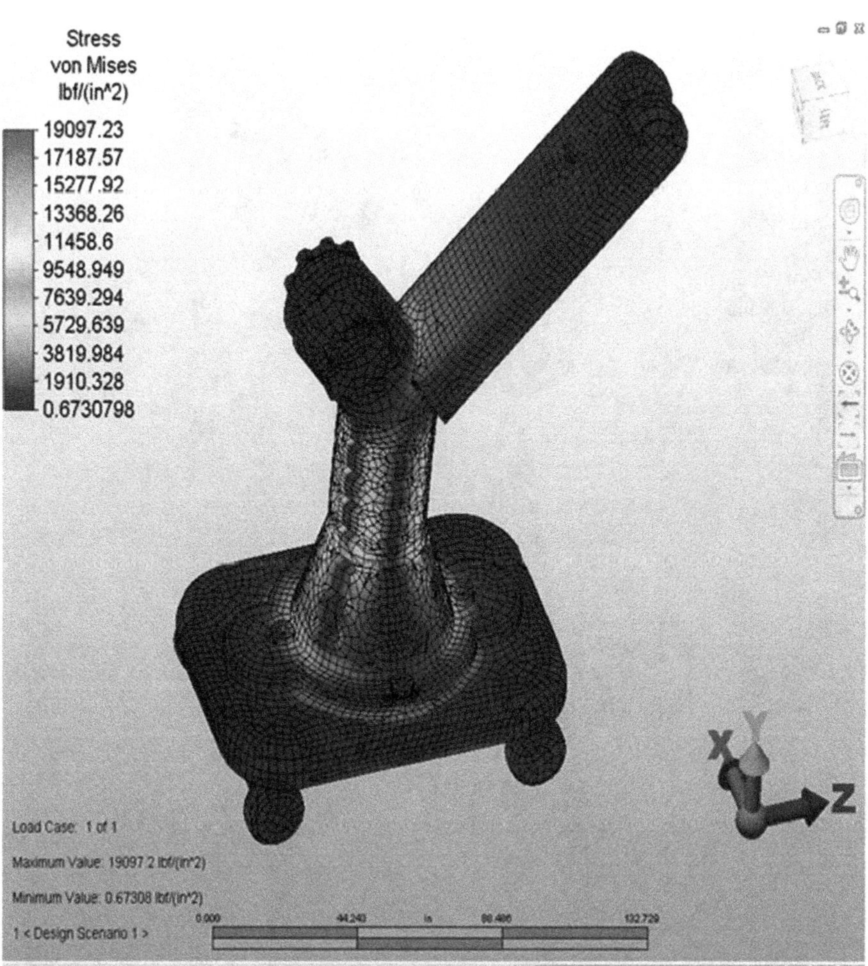

Fig. 7.18 Computer-aided simulation of stress profile in the base of new robotic system

Table 7.6 shows that while the robot arm of this new robotic system moves to the lower direction, the prototype testing results of PWM register prove the normal function as the testing values are almost same as the simulated PWM register results by computer-aided modeling and simulation as shown in Fig. 7.12.

The kinematic motion of prototyped robotic system traced and detected by PWM registers shows 3D movement of robot arm in this new robotic system. The prototype experiment indicates that the robot arm moves in 3D directions with 8.02 ft in the direction of front to back, 12.05 ft in the direction of left to right, and 10.04 ft in the direction of top to bottom. Comparing to computer-aided simulation of 8 ft in front-to-back direction, 12 ft in left-to-right direction, and 10 ft in top-to-bottom direction previously shown in Fig. 7.13, the experimental results are very close to the simulated results.

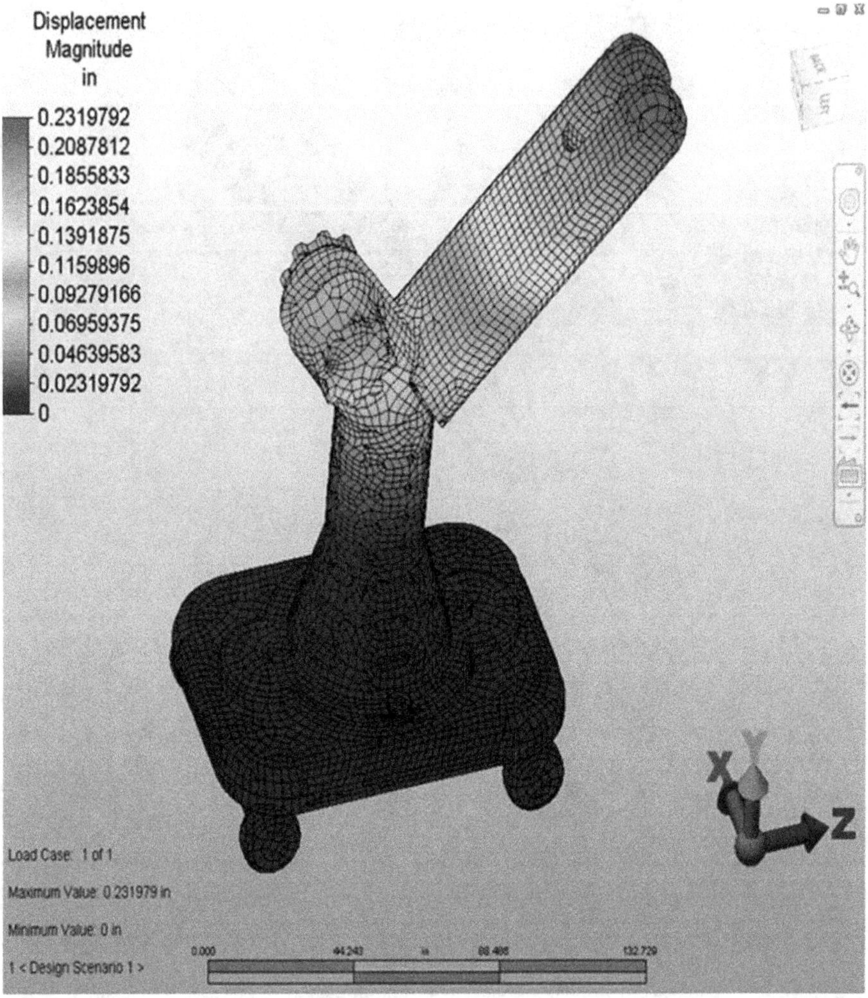

Fig. 7.19 Computer-aided simulation of deflection profile in the base of new robotic system

Table 7.1 Prototype experimental results of PWM register as robot arm moving to the front

Time (ms)	PWM register 1	PWM register 2
0.5	0	0
1.0	178	−178
1.5	305	−305
2.0	410	−410
2.5	425	−425
3.0	468	−468
3.5	585	−585
4.0	648	−648
4.5	725	−725

Table 7.2 Prototype experimental results of PWM register as robot arm moving to the back

Time (ms)	PWM register 1	PWM register 2
0.5	0	0
1.0	188	−188
1.5	378	−378
2.0	405	−405
2.5	415	−415
3.0	448	−448
3.5	605	−605
4.0	708	−708
4.5	745	−745

Table 7.3 Prototype experimental results of PWM register as robot arm moving to the left

Time (ms)	PWM register 1	PWM register 2
0.5	0	0
1.0	−248	−450
1.5	−380	−375
2.0	−625	−980
2.5	−780	−1,218
3.0	−578	−915
3.5	−775	−1,420
4.0	−978	−1,550
4.5	−1065	−1,725

Table 7.4 Prototype experimental results of PWM register as robot arm moving to the right

Time (ms)	PWM register 1	PWM register 2
0.5	0	0
1.0	455	245
1.5	480	475
2.0	878	705
2.5	685	490
3.0	675	380
3.5	948	725
4.0	1,230	875
4.5	1,378	1,025

The experiment has also been performed to assert if this newly developed robotic system has required well-set structure to handle different industrial applications including automated and high-speed manufacturing and production. The structure and strength testing have been carried out on some critical robotic components with stress and deflection measurement displayed in the following tables.

Table 7.7 demonstrates the experimental results of maximum stress and maximum deflection of robot arm in this new robotic system.

The prototype testing results of robot arm in Table 7.7 prove the normal function as the average maximum stress 22,421.56 psi and average maximum deflection 0.0110 in. are very close to the results of maximum stress 22,421.51 psi and

Table 7.5 Prototype testing results of PWM register as robot arm moving to the upper direction

Time (ms)	PWM register
0.5	0
1.0	248
1.5	495
2.0	625
2.5	755
3.0	945
3.5	1,188
4.0	1,375
4.5	1,695

Table 7.6 Prototype testing results of PWM register as robot arm moving to the lower direction

Time (ms)	PWM register
0.5	0
1.0	−255
1.5	−635
2.0	−818
2.5	−1,050
3.0	−1,245
3.5	−1,375
4.0	−1,620
4.5	−1,705

Table 7.7 Prototype testing results of maximum stress and maximum deflection of robot arm in this new robotic system

Number of experiment	Maximum stress (psi)	Maximum deflection (in.)
1	22,421.57	0.0108
2	22,421.54	0.0104
3	22,421.48	0.0102
4	22,421.49	0.0115
5	22,421.62	0.0115
6	22,421.61	0.0104
7	22,421.56	0.0113
8	22,421.59	0.0103
9	22,421.49	0.0113
10	22,421.62	0.0109
11	22,421.61	0.0111
12	22,421.55	0.0112
13	22,421.57	0.0103
14	22,421.48	0.0115
15	22,421.63	0.0113
16	22,421.61	0.0112
17	22,421.57	0.0101
18	22,421.59	0.0115
19	22,421.57	0.0114
20	22,421.48	0.0108
Average	22,421.56	0.0110

Table 7.8 Prototype experimental results of maximum stress and maximum deflection of robot arm link in this new robotic system

Number of experiment	Maximum stress (psi)	Maximum deflection (in.)
1	22,634.91	0.0455
2	22,634.93	0.0445
3	22,634.88	0.0448
4	22,634.98	0.0456
5	22,634.95	0.0457
6	22,634.89	0.0454
7	22,634.86	0.0446
8	22,634.92	0.0454
9	22,634.91	0.0456
10	22,634.90	0.0449
11	22,634.88	0.0445
12	22,634.98	0.0457
13	22,634.91	0.0455
14	22,634.93	0.0446
15	22,634.99	0.0457
16	22,634.92	0.0455
17	22,634.98	0.0443
18	22,634.93	0.0455
19	22,634.90	0.0453
20	22,634.98	0.0445
Average	22,634.93	0.0452

maximum deflection 0.0105 in. that are presented, respectively, in Figs. 7.14 and 7.15 by computer-aided modeling and numerical simulation.

Table 7.8 demonstrates the experimental results of maximum stress and maximum deflection of robot arm link in this new robotic system.

The prototype experimental results of robot arm link in Table 7.8 verify the proper function because the average maximum stress 22,634.93 psi and average maximum deflection 0.0452 in. are closely equal to the results of maximum stress 22,634.97 psi and maximum deflection 0.0447 in. that are represented, respectively, in Figs. 7.16 and 7.17 by computer-aided modeling and numerical simulation.

Table 7.9 represents the experimental results of maximum stress and maximum deflection of robot arm base in this new robotic system.

The prototype experimental results of robot arm base in Table 7.9 confirm the appropriate function since the average maximum stress 19,097.27 psi and average maximum deflection 0.2316 in. are approximately same as the results of maximum stress 19,097.23 psi and maximum deflection 0.2320 in. that are laid out, respectively, in Figs. 7.18 and 7.19 by computer-aided modeling and numerical simulation.

Table 7.9 Prototype experimental results of maximum stress and maximum deflection of robot arm base in this new robotic system

Number of experiment	Maximum stress (psi)	Maximum deflection (in.)
1	19,097.27	0.2318
2	19,097.29	0.2322
3	19,097.32	0.2311
4	19,097.35	0.2310
5	19,097.36	0.2312
6	19,097.21	0.2318
7	19,097.22	0.2315
8	19,097.38	0.2312
9	19,097.27	0.2323
10	19,097.25	0.2311
11	19,097.29	0.2314
12	19,097.18	0.2313
13	19,097.20	0.2322
14	19,097.27	0.2312
15	19,097.35	0.2315
16	19,097.27	0.2313
17	19,097.29	0.2320
18	19,097.20	0.2318
19	19,097.21	0.2312
20	19,097.25	0.2320
Average	19,097.27	0.2316

7.4 Discussion and Future Improvement on Robotic System

To date, there is considerable evidence that more and more product companies are applying robotic systems to their scope of production practice due to significant benefits for the business but some firms are little concerned about the cost. The robotic technology is exceptional with 3D manipulated visual images and systems can be oriented intuitively with degrees of freedom in multiple directions. In addition, the robotic stability that the technicians fully control can significantly reduce the unexpected defects caused by human errors and robotic technology largely justifies the technical limitation to avoid the deficiencies of normal production technique. However, there is still something required to be improved for this new robotic system. For example, it remains difficult-to-operate system in multi-quadrants required for some complicated production processes. Also, the robotic system cost needs to be decreased. To make these improvements, the current robotic system modification should include modifying robotic system design for cost-effective and easy maintenance, improving system design for more reliable function, and studying optimal robotic mechanism through computer-aided simulation for high energy and mechanical efficiency. More prototype experiments have been planned to implement this new robotic system and the prototype will be also sent to fields for further evaluations.

Magnetic Sealing System 8

8.1 Design of Magnetic Sealing System

The gas and lubrication oil leakages in reciprocating machinery systems including compressors and cooling units are usual problems that have not been well solved yet which normally influences the machine performance. This new magnetic sealing mechanism design is to solve these gaseous or fluid media leaking problems. In this new sealing mechanism, the rare-earth magnet steel can perform as a permanent magnet. Both computer-aided simulation and prototype testing verify that this sealing mechanism can effectively decrease the leaking problem in machinery systems including compressors and cooling systems. This new mechanism has better sealing performance than traditional rubber seal, diaphragm seal, corrugated pipe seal, and magnetic fluid seal. Figures 8.1, 8.2, and 8.3 display full view of new high-pressure gas compressive and filtration system, 3D cross-section view of new high-pressure gas compressive system, and 2D cross-section view of new high-pressure gas compressive system.

As shown in Figs. 8.2 and 8.3, the rare-earth magnet steel can perform as a permanent magnet steel with very higher density of magnetic flux Br, strong magnetic field H_g, and larger product of magnetism and energy $(BH)_{large}$ that keeps magnetic particles to be strongly bonded to the inner wall of permanent magnet steel. The main advanced features of this magnetic sealing system include raised Br in the working gap of magnetic circuit, more durable function in sealing performance, longer lifetime in sealing capability, more compact in system configuration, light in unit weight, higher in efficiency, and more reliable in sealing function. When machine piston moves back and forth inside the cylinder, the lubricating oil is sealed by intensive magnetic particles that are firmly bonded in the inner wall surface of magnet steel as lubricating oil move toward the seal. As soon as lubrication oil has been blocked by magnetic particles, the oil droplets will drop by its gravity to the bottom container in main crankshaft chamber that can keep lubricating oil in crankshaft chamber from entering the compressive gas cylinder. Also the gas leaking can be significantly reduced because it is very

J. Zheng Li, *CAD, 3D Modeling, Engineering Analysis, and Prototype Experimentation*, DOI 10.1007/978-3-319-05921-1_8

Fig. 8.1 Full view of new high-pressure gas compressive and filtration system

Fig. 8.2 3D cross-section view of new high-pressure gas compressive system

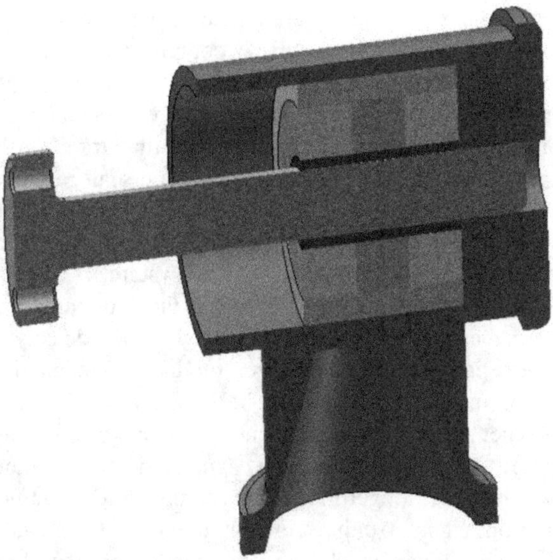

difficult for gas to penetrate through the firmly bonded magnetic particle layers. To design and develop this new magnetic sealing system, two important factors that must be taken account in order to maintain its reliable function are magnetic flux density and material magnetic stability. Therefore the magnetic flux in the magnetic circuit of this sealing mechanism should be kept reliable over a long time period and material magnetic field should be extremely stable enough to handle the external magnetic field variation, temperature fluctuation, mechanical vibration, unexpected shock, and severe environmental changes. The higher magnetic flux density Br, higher magnetic field intensity H_g, and larger product of magnetism and energy $(BH)_{large}$ in magnetic circuit design are targeted to maintain their peak values in this new magnetic sealing system.

8.2 Computer-Aided Simulation on Magnetic Sealing System

The magnetic circuit in this sealing mechanism can be considered in static condition so that the ampere enclosed circuit and *H–B* curve are able to be used to analyze the functionality of this rare-earth magnet steel. The magnetic circuit in this magnetic sealing system can be assumed as a series magnetic circuit mainly consisting of magnet steel and sealing gap. The following equations can be derived based on Fig. 8.3 (Ekren et al. 2011):

$$H \times L + H_g \times L_g = 0 \tag{8.1}$$

$$H \times L = -\frac{L_g \times \Phi}{U_0 \times A_g} \tag{8.2}$$

The magnetic curve in magnetic circuit is shown in Fig. 8.4.

If $\mathrm{Fm}(\Phi) = H \times L$, the intersection point between function $\{\mathrm{Fm}(\Phi)\}$ and line of $\{-[L_g/(U_0 \times A_g)] \times \Phi\}$ at vertical coordinate system is the magnetic flux in sealing gap that needs to be determined. The sealing gap reduces from L_g to L_g' if more magnetic particles are put into the gap of magnetic circuit. As the thickness of magnetic particle layer in the gap between contact surfaces of magnet steel and shaft to be sealed varies from 0 to b, the functioning point of magnet steel varies

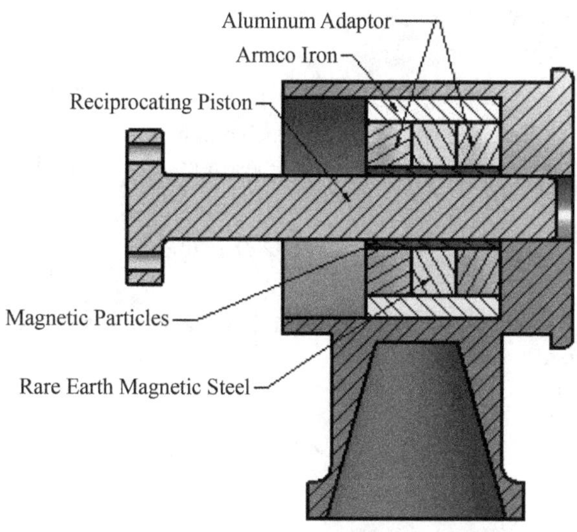

Fig. 8.3 2D cross-section view of new high-pressure gas compressive system

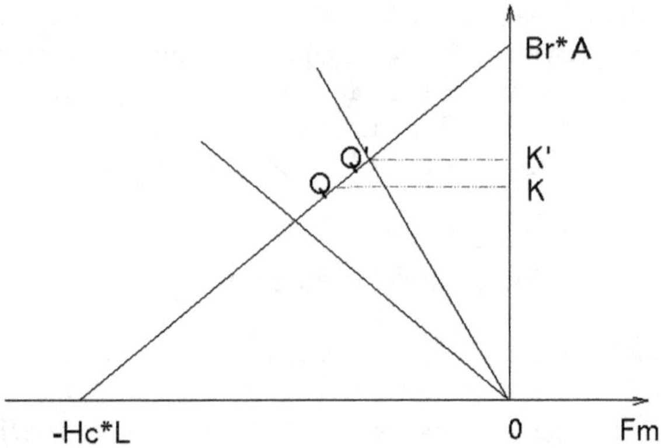

Fig. 8.4 Magnetic curve of circuit

along the line QK and magnetic flux in sealing gap can be solved. The coefficient of magnetic efficiency η_{eff} can be applied to verify if the magnetic field in sealing mechanism is correctly designed (Evans et al. 2006):

$$\eta_{\mathrm{eff}} = \frac{{B_{\mathrm{g}}}^2 \times V_{\mathrm{g}}}{(B \times H)_{\max} \times V} \tag{8.3}$$

A higher η_{eff} value shows more reliable magnetic circuit design. The normal η_{eff} value is around 40 % in standard spec. The computational simulation indicates that the η_{eff} value in this new magnetic sealing mechanism is 48.8 % which confirms the proper magnetic circuit design in this newly developed magnetic sealing system. Figure 8.5 shows the cross-section view of magnetic steel in this sealing system. The mathematical equation of calculating sealing capacity ΔP can be derived from energy balance law (Feil-Seifer et al. 2007).

Based on diagram in Fig. 8.5, $R_1 = \frac{2 \times b}{\sin(\alpha)}$, $R_2 = \frac{2 \times b}{\sin(\beta)}$, $S_1 = R_1 \times \alpha$, and $S_2 = R_2 \times \beta$.

So,

$\Delta S = S_2 - S_1 = \left[\frac{\beta}{\sin(\beta)} - \frac{\alpha}{\sin(\alpha)}\right] \times 2 \times b$ and $\underline{OO'} = 2 \times \beta \times [ctg(\alpha) - ctg(\beta)]$.

Since the work done by each magnetic force line equals to $\{T \times \Delta S\}$, total work done by all magnetic force lines in magnetic circuit is

$$W_{\mathrm{ML}} = B \times D \times T \times 2b \times \left[\frac{\beta}{\sin(\beta)} - \frac{\alpha}{\sin(\alpha)}\right] \tag{8.4}$$

And the work done by gaseous media pressure exerted to the body of magnetic particles is

$$W_{\mathrm{GP}} = 4 \times b^2 \times \Delta P \times \left[\frac{\sin(\alpha) + \pi \times [\cos(\alpha) - 2]^2}{8 \times \sin(\alpha) - \frac{\sin(\alpha)}{4}}\right] \tag{8.5}$$

Fig. 8.5 Cross-section view of magnetic steel

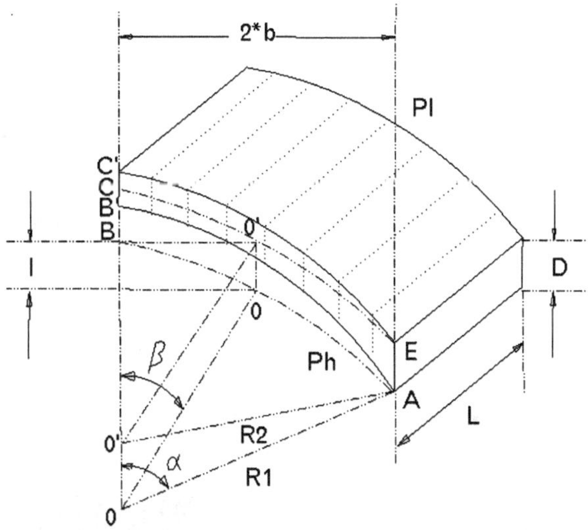

Based on the energy balance law, the work done by magnetic force lines applied in magnetic circuit should equal to the work done by gaseous media pressure applied to the body of magnetic particles. So,

$$\Delta \mathrm{P} = \frac{\left(\frac{1}{2}\right) \times B \times D \times T \times \left[\sin(\alpha) - \alpha\right]}{\left\{2 \times b \times (\sin \alpha) + \frac{\alpha}{8} \times \left[\cos^2(\alpha) - 4 \times \cos(\alpha) + 3\right]\right\}} \tag{8.6}$$

The above equation can be analyzed using computer-aided modeling and numerical simulation. The stress and deflection profiles of major components in this new magnetic sealing system are shown in Figs. 8.6, 8.7, 8.8, 8.9, 8.10, 8.11, 8.12, 8.13, 8.14, 8.15, 8.16, 8.17, 8.18, 8.19, and 8.20.

The computer-aided simulation and analysis in Figs. 8.7 and 8.8 display the stress and deflection of aluminum adaptor in this new magnetic sealing system. The analytic results show that the maximum stress of 2,923.21 psi in this aluminum adaptor is less than the material yield strength and maximum deflection of 0.000016 in. is within material allowable deflection limit.

The computer-aided simulation and analysis in Figs. 8.10 and 8.11 tell the stress and deflection of Armco iron ring in this new magnetic sealing system. The analytic results present that the maximum stress of 2,428.89 psi in this Armco iron ring is less than the material yield strength and maximum deflection of 0.00001 in. is within material allowable deflection limit.

The computer-aided simulation and analysis in Figs. 8.13 and 8.14 demonstrate the stress and deflection of compressive unit in this new magnetic sealing system. The analytic results indicate that the maximum stress of 19,376.24 psi in this compressive unit is less than the material yield strength of 36,000 psi and maximum deflection of 0.00105 in. is within material allowable deflection limit.

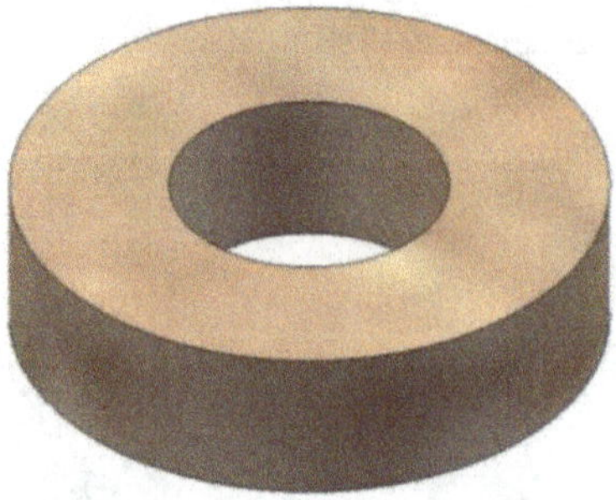

Fig. 8.6 Aluminum adaptor

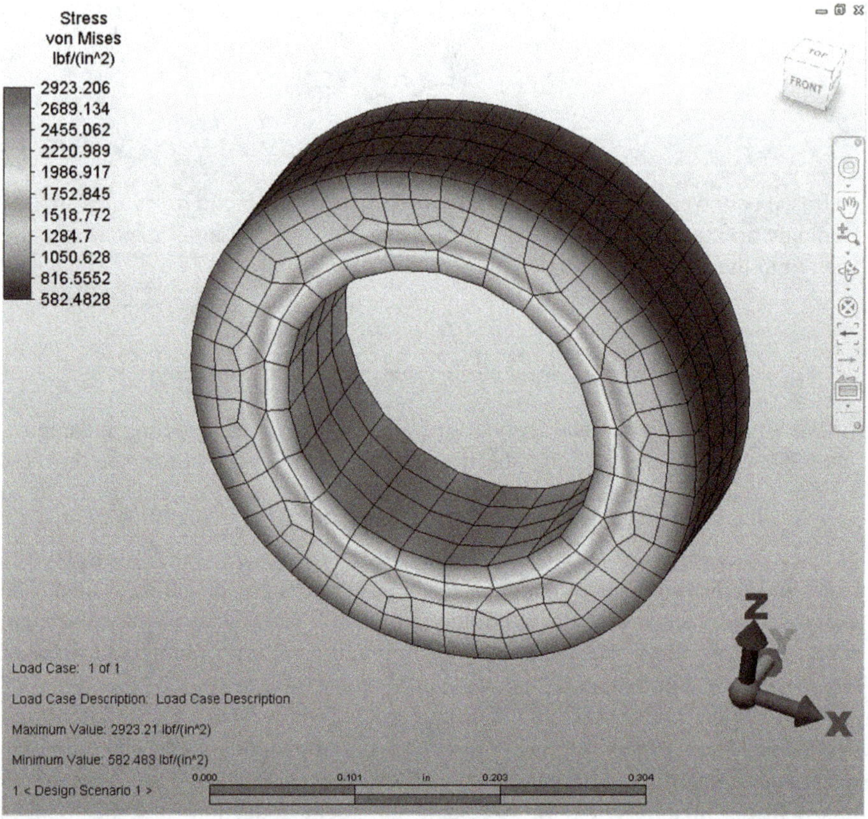

Fig. 8.7 Stress profile in aluminum adaptor

The computer-aided simulation and analysis in Figs. 8.16 and 8.17 state the stress and deflection of piston shaft in this new magnetic sealing system. The analytic results display that the maximum stress of 19,484.07 psi in this piston shaft is less than the material yield strength of 36,000 psi and maximum deflection of 0.00021 in. is within material allowable deflection limit.

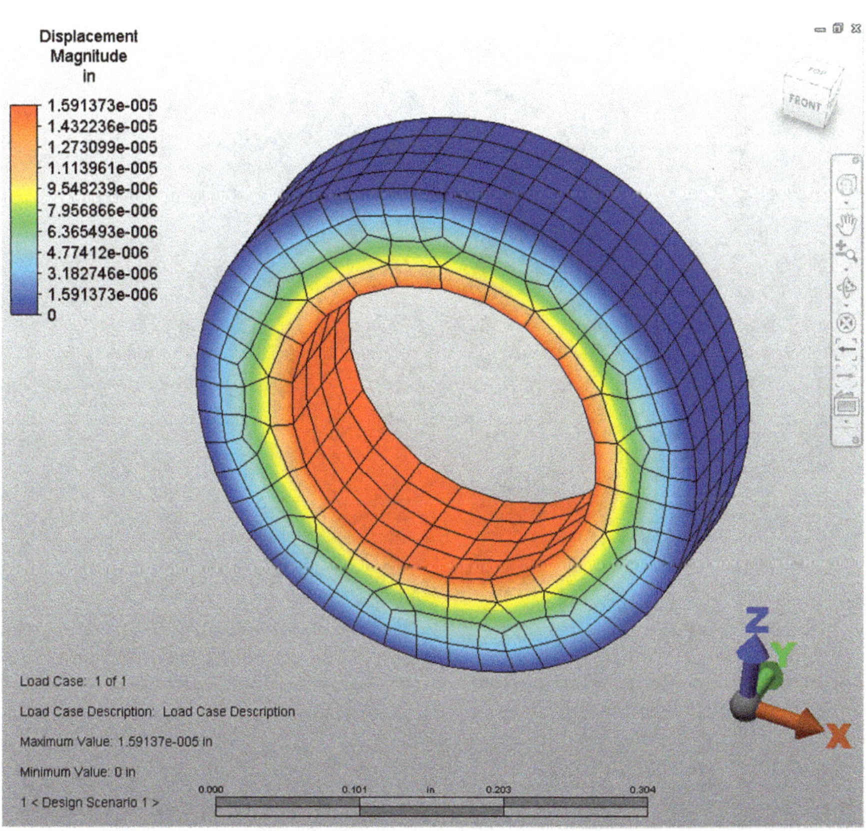

Fig. 8.8 Deflection profile in aluminum adaptor

Fig. 8.9 Armco iron ring

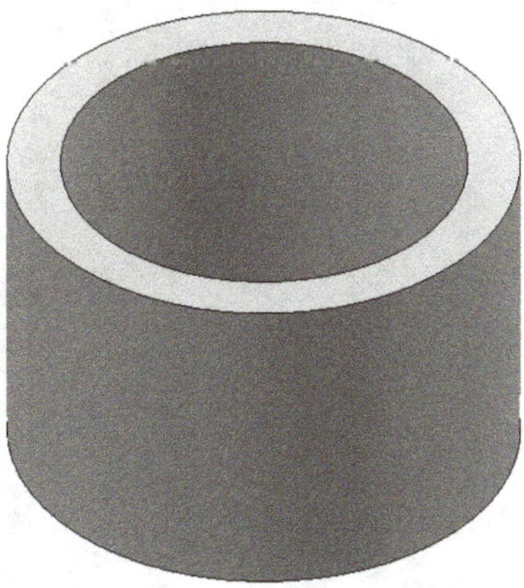

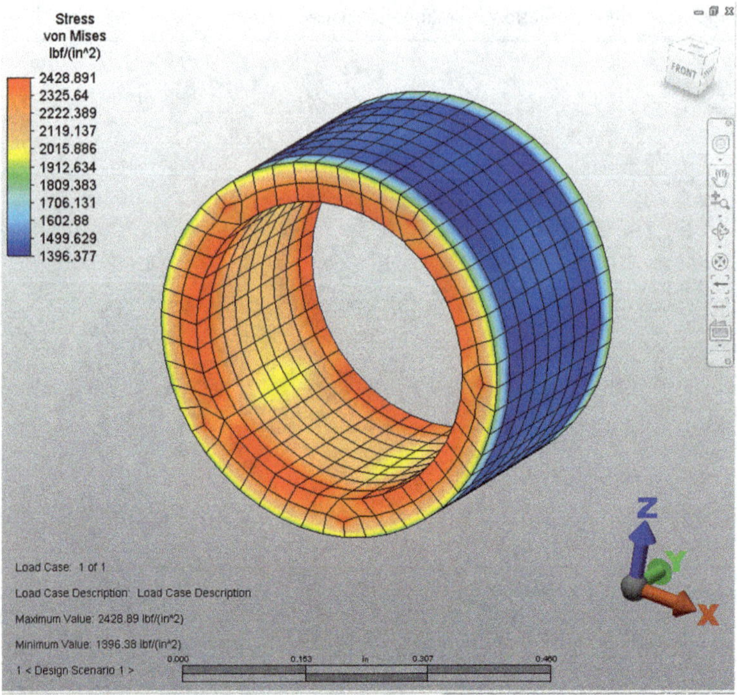

Fig. 8.10 Stress profile in Armco iron ring

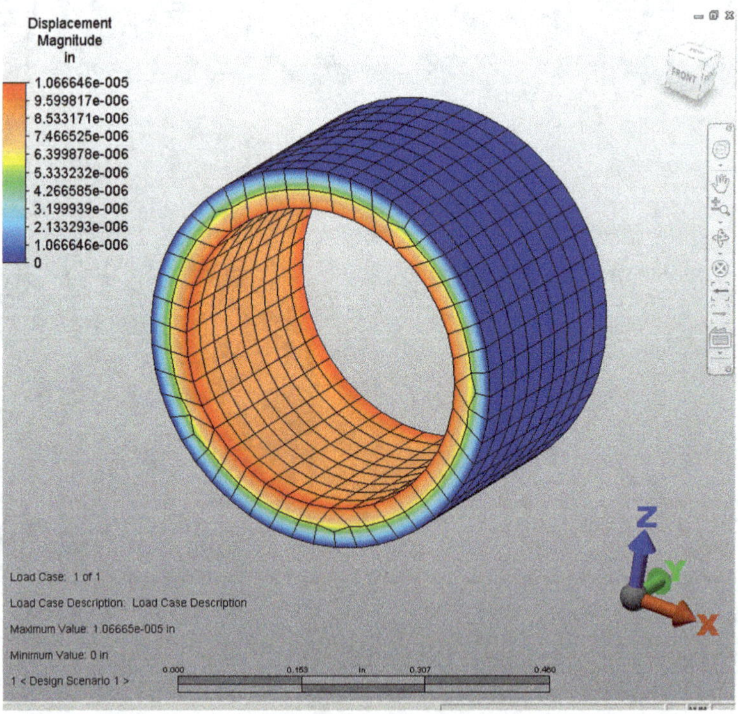

Fig. 8.11 Deflection profile in Armco iron ring

Fig. 8.12 Compressive unit

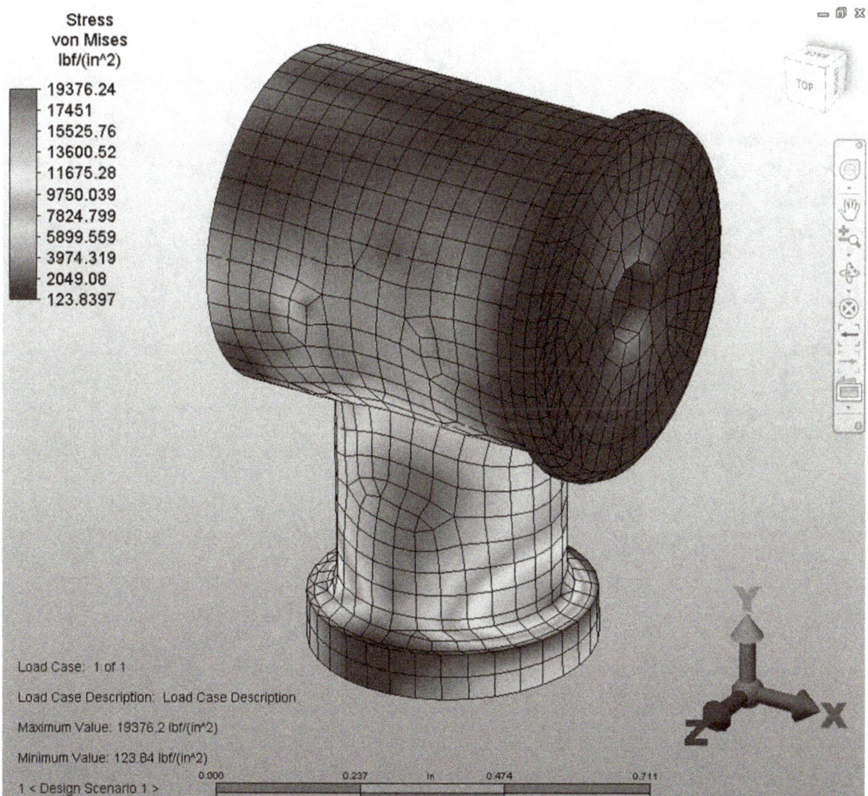

Fig. 8.13 Stress profile in compressive unit

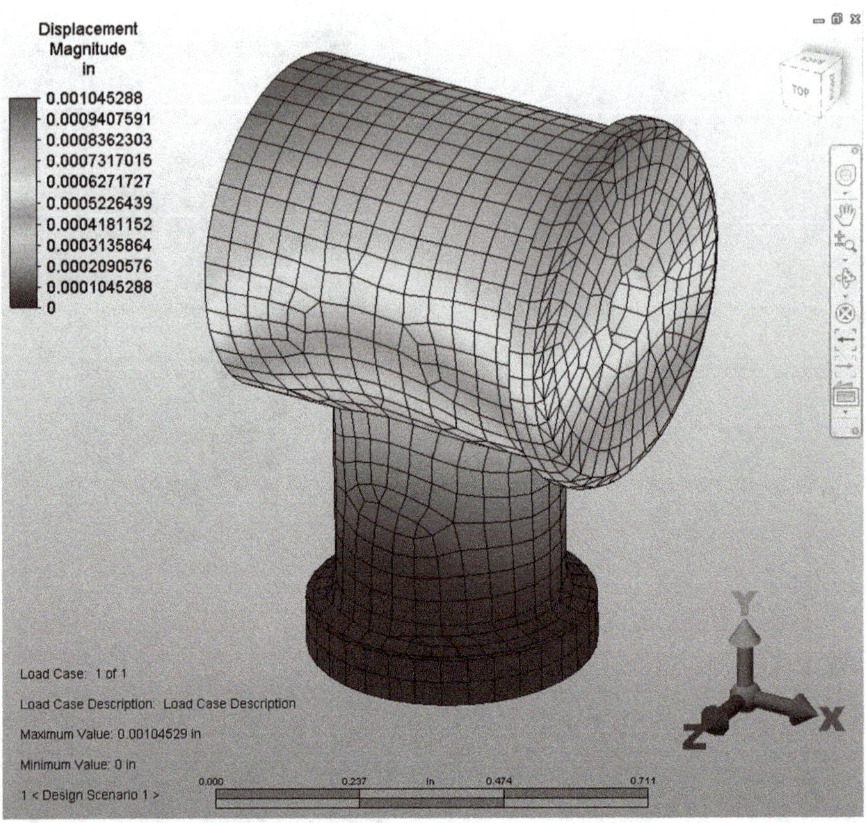

Fig. 8.14 Deflection profile in compressive unit

Fig. 8.15 Piston shaft

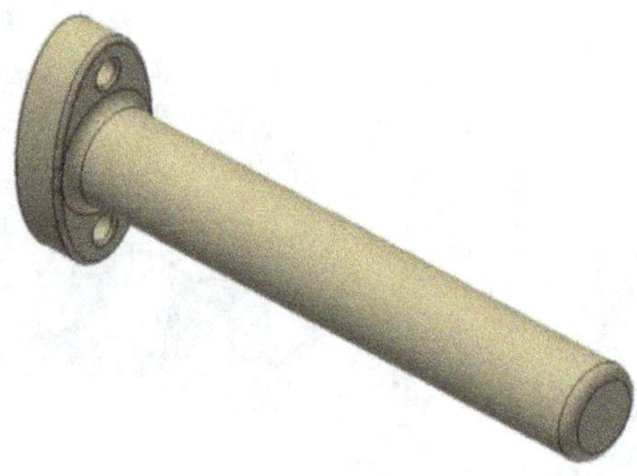

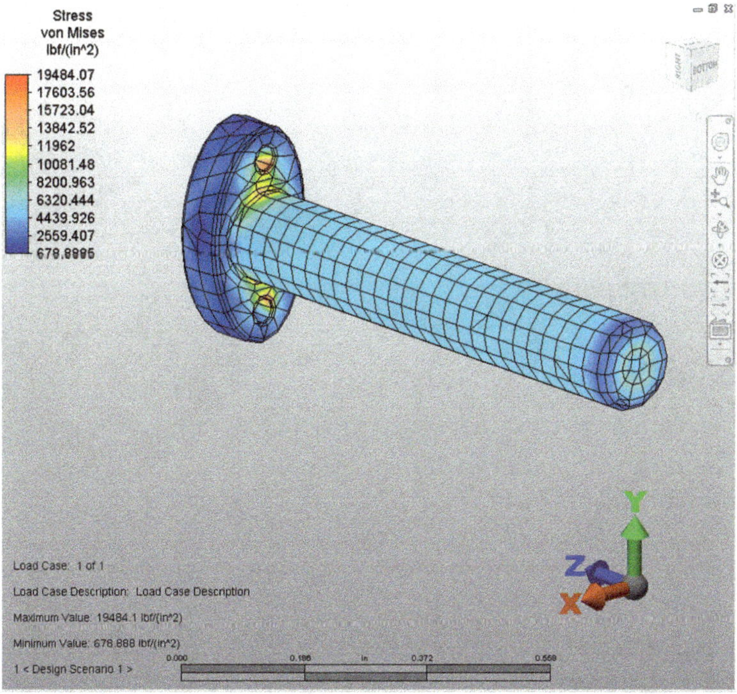

Fig. 8.16 Stress profile in piston shaft

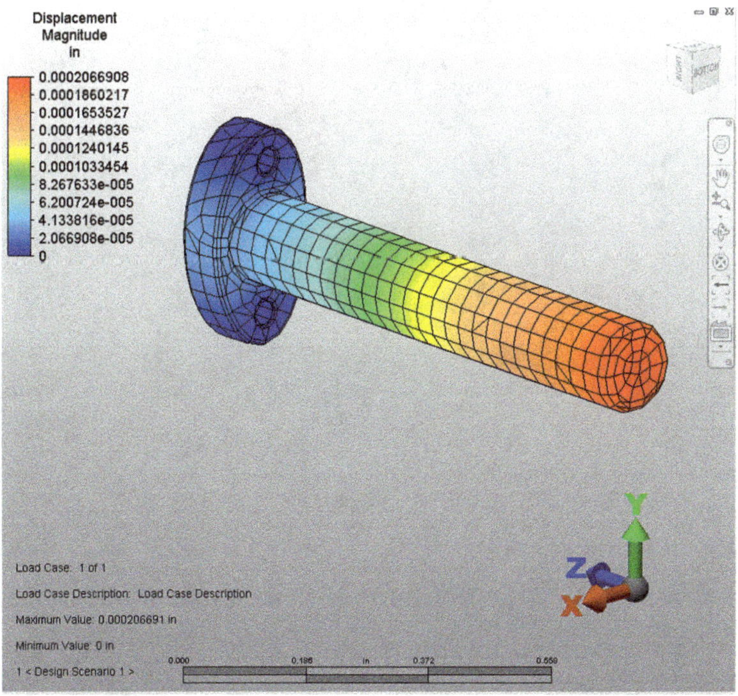

Fig. 8.17 Deflection profile in piston shaft

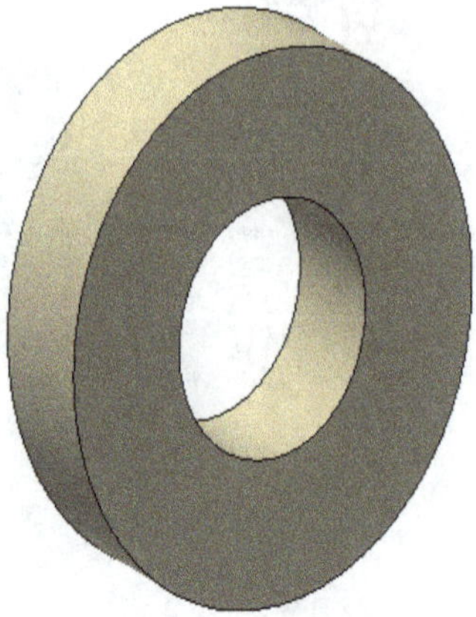

Fig. 8.18 Rare-earth magnetic steel

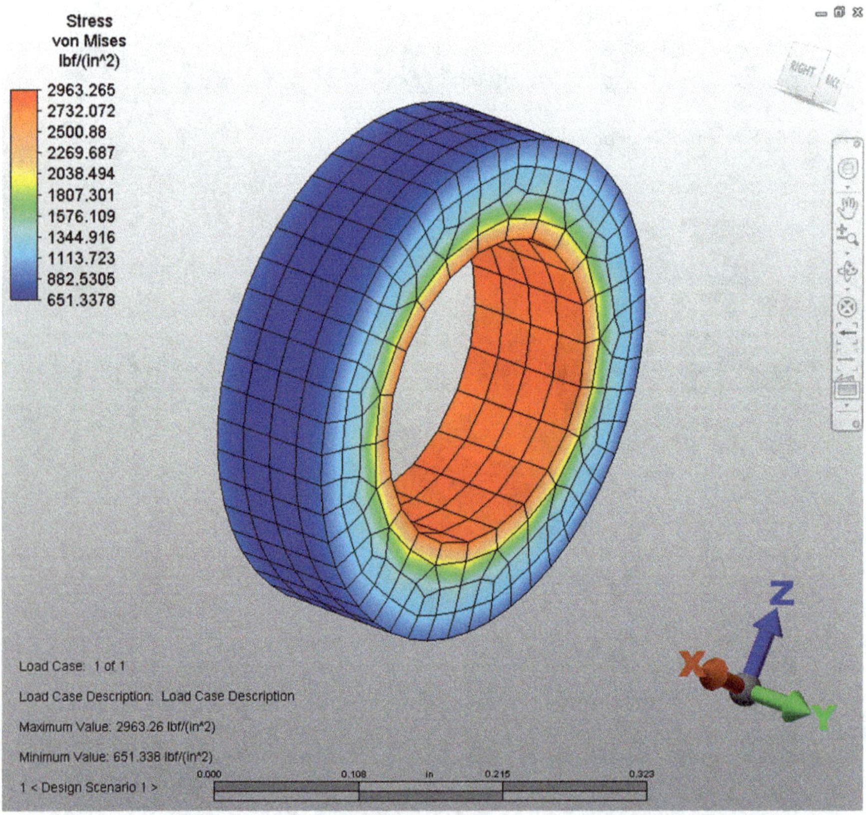

Fig. 8.19 Stress profile in rare-earth magnetic steel

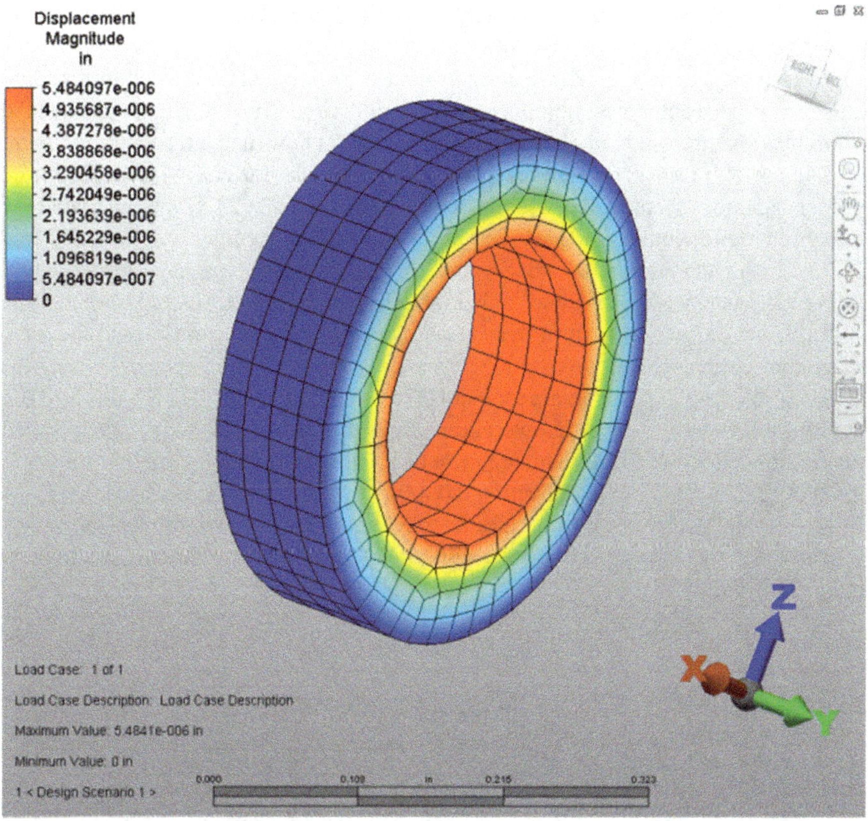

Fig. 8.20 Deflection profile in rare-earth magnetic steel

The computer-aided simulation and analysis in Figs. 8.19 and 8.20 show the stress and deflection of rare-earth magnetic steel in this new magnetic sealing system. The analytic results tell that the maximum stress of 2,963.27 psi in this rare earth magnetic steel is less than the material yield strength and maximum deflection of 0.000005 in. is within material allowable deflection limit.

These above computational simulation results demonstrate that the maximum stresses on these major components are all less than the material yield stress and maximum material deflections are all within material allowable deformation limits. The seal capacity of this new magnetic sealing system can keep the oil leakage from chamber of crankshaft into the gaseous chamber of cooling system and reciprocating machineries. The computational solutions verify that this newly developed magnetic sealing system can work properly for better sealing functions.

8.3 Experiment on Magnetic Sealing System

The newly designed magnetic sealing system is prototyped and tested to compare and verify the results from computer-aided simulation. Table 8.1 demonstrates the prototype testing results of aluminum adaptor in this new magnetic sealing system.

The prototype experimental results of aluminum adaptor in Table 8.1 verify the proper function because the average maximum stress 2,923.37 psi and average maximum deflection 0.00002 in. are very similar to the results of maximum stress 2,923.21 psi and maximum deflection 0.00001 in. that are represented, respectively, in Figs. 8.7 and 8.8 by computer-aided modeling and numerical simulation.

Table 8.2 expresses the prototype testing results of Armco iron ring in this new magnetic sealing system.

The prototype experimental results of Armco iron ring in Table 8.2 confirm the appropriate function since the average maximum stress 2,428.77 psi and average maximum deflection 0.00002 in. are closely equal to the results of maximum stress 2,428.89 psi and maximum deflection 0.00001 in. that are indicated, respectively, in Figs. 8.10 and 8.11 by computer-aided modeling and numerical simulation.

Table 8.3 records the prototype testing results of compressive unit in this new magnetic sealing system.

Table 8.1 Prototype testing of aluminum adaptor in this new magnetic sealing system

Number of experiment	Maximum stress (psi)	Maximum deflection (in.)
1	2,923.24	0.00002
2	2,923.18	0.00001
3	2,923.27	0.00001
4	2,923.35	0.00002
5	2,923.38	0.00001
6	2,923.44	0.00002
7	2,923.49	0.00001
8	2,923.48	0.00002
9	2,923.47	0.00002
10	2,923.36	0.00001
11	2,923.24	0.00001
12	2,923.29	0.00001
13	2,923.37	0.00001
14	2,923.39	0.00002
15	2,923.47	0.00003
16	2,923.38	0.00001
17	2,923.48	0.00002
18	2,923.45	0.00001
19	2,923.38	0.00001
20	2,923.35	0.00002
Average	2,923.37	0.00002

Table 8.2 Prototype testing of Armco iron ring in this new magnetic sealing system

Number of experiment	Maximum stress (psi)	Maximum deflection (in.)
1	2,428.89	0.00002
2	2,428.89	0.00001
3	2,428.89	0.00001
4	2,428.88	0.00002
5	2,428.91	0.00002
6	2,428.84	0.00001
7	2,428.78	0.00001
8	2,428.72	0.00001
9	2,428.68	0.00002
10	2,428.69	0.00001
11	2,428.59	0.00002
12	2,428.75	0.00002
13	2,428.72	0.00002
14	2,428.71	0.00001
15	2,428.73	0.00002
16	2,428.75	0.00002
17	2,428.78	0.00002
18	2,428.77	0.00001
19	2,428.75	0.00002
20	2,428.72	0.00001
Average	2,428.77	0.00002

Table 8.3 Prototype testing of compressive unit in this new magnetic sealing system

Number of experiment	Maximum stress (psi)	Maximum deflection (in.)
1	19,376.25	0.00108
2	19,376.24	0.00112
3	19,376.38	0.00104
4	19,376.39	0.00109
5	19,376.36	0.00112
6	19,376.54	0.00118
7	19,376.39	0.00115
8	19,376.37	0.00112
9	19,376.38	0.00108
10	19,376.48	0.00112
11	19,376.45	0.00111
12	19,376.37	0.00115
13	19,376.42	0.00117
14	19,376.39	0.00118
15	19,376.45	0.00116
16	19,376.38	0.00112
17	19,376.54	0.00102
18	19,376.45	0.00116
19	19,376.41	0.00112
20	19,376.38	0.00110
Average	19,376.40	0.00112

Table 8.4 Prototype testing of piston shaft in this new magnetic sealing system

Number of experiment	Maximum stress (psi)	Maximum deflection (in.)
1	19,484.12	0.00019
2	19,484.08	0.00020
3	19,484.11	0.00017
4	19,484.15	0.00015
5	19,484.18	0.00018
6	19,484.05	0.00016
7	19,484.16	0.00012
8	19,484.24	0.00013
9	19,484.29	0.00015
10	19,484.33	0.00017
11	19,484.38	0.00014
12	19,484.39	0.00013
13	19,484.35	0.00012
14	19,484.29	0.00015
15	19,484.24	0.00012
16	19,484.18	0.00011
17	19,484.26	0.00022
18	19,484.19	0.00018
19	19,484.17	0.00013
20	19,484.21	0.00014
Average	19,484.22	0.00015

The prototype experimental results of compressive unit in Table 8.3 prove the normal function as the average maximum stress 19,376.40 psi and average maximum deflection 0.00112 in. are almost same as the results of maximum stress 19,376.24 psi and maximum deflection 0.00105 in. that are laid out, respectively, in Figs. 8.13 and 8.14 by computer-aided modeling and numerical simulation.

Table 8.4 states the prototype testing results of piston shaft in this new magnetic sealing system.

The prototype experimental results of piston shaft in Table 8.4 verify the proper function because the average maximum stress 19,484.22 psi and average maximum deflection 0.00015 in. are approximately equal to the results of maximum stress 19,484.07 psi and maximum deflection 0.00021 in. that are shown, respectively, in Figs. 8.16 and 8.17 by computer-aided modeling and numerical simulation.

Table 8.5 demonstrates the prototype testing results of rare-earth magnetic steel in this new magnetic sealing system.

The prototype experimental results of rare-earth magnetic steel in Table 8.5 confirm the appropriate function as the average maximum stress 2,963.16 psi and average maximum deflection 0.000012 in. are approximately same as the results of maximum stress 2,963.27 psi and maximum deflection 0.000006 in. that are presented, respectively, in Figs. 8.19 and 8.20 by computer-aided modeling and numerical simulation.

Table 8.5 Prototype testing of rare-earth magnetic steel in this new magnetic sealing system

Number of experiment	Maximum stress (psi)	Maximum deflection (in.)
1	2,963.24	0.000008
2	2,963.21	0.000004
3	2,963.18	0.000009
4	2,963.15	0.000011
5	2,963.21	0.000012
6	2,963.13	0.000010
7	2,963.14	0.000009
8	2,963.16	0.000014
9	2,963.21	0.000018
10	2,963.18	0.000015
11	2,963.12	0.000016
12	2,963.14	0.000013
13	2,963.15	0.000012
14	2,963.12	0.000015
15	2,963.15	0.000011
16	2,963.12	0.000012
17	2,963.21	0.000015
18	2,963.12	0.000012
19	2,963.11	0.000008
20	2,963.19	0.000009
Average	2,963.16	0.000012

8.4 Discussion and Future Improvement on Magnetic Sealing System

Good sealing system can prevent liquid or gas from leaking in machinery to improve performance and increase efficiency. A new magnetic sealing system has been designed and developed with magnetic particles embedded in a working gap between external rotational or reciprocating shaft and internal wall of rare-earth magnetic steel. The magnetic circuit formed on the internal surface of magnetic steel and external surface of rotational or reciprocating can dynamically seal the working gap to avoid the lubrication oil or gaseous media leaks. In this new magnetic sealing system design, the rotational or the reciprocating shafts can rotate or reciprocate while being supported in a metal non-contact style which reduces the frictional force significantly. Currently this new magnetic sealing system demonstrates its strong dynamic sealing capability in rotating or reciprocating machinery. The size of sealing unit is decreased since the components in sealing unit are reduced. Future improvement will include modifying the working gap to determine what geometry and dimension will get more intensified magnetic flux in the magnetic circuit, performing more computational simulations to optimize the sealing unit design, and investigating different materials for unit cost reduction.

Automated and High-Speed Packaging System 9

9.1 Design of Automated and High-Speed Packaging Machinery System

The automated packing machinery becomes more popular in many different industries to reduce labor cost. This new packaging machinery system can automatically load carton papers, open carton paper to the square box, seal the bottom flap, seal top flap after putting the products into box, label print, and offload the packing box to the loading area in the production line. Figure 9.1 shows the layout of this new automated packaging assembly system. The details of full packaging production sequence of this new packaging system are described as follows:

The carton papers are loaded into the loading channel and each carton paper is sucked into the production inline from loading channel by vacuum cup. There is a support plate at the front of vacuum cup and support plate is aligned with carton paper. The carton paper falls down before vacuum cup picks it up. The soft spring leaf in support plate can keep carton paper fixed while opening square box. At each side of box bottom, there are two movable pushing plates that first move up to flatten two box short flaps and flatten two long flaps thereafter. Two pushing plates move to the left or the right separately and the box bottom is finally sealed by bottom sealing mechanism. After sealing the box bottom, a pushing plate at the left side will drive box to the next station to print label at two sides of box. A support plate at the top of packing box is to make sure that the labels printed at two sides are horizontal with no titled angle. To close and seal the top flaps of packing box after loading the products, three mechanisms can be introduced including divider, pusher, and final curvature. Before packing box moving to contact final curvature, the divider separates long flaps of box top into two sides to keep them staying with the wall with no contact with short flaps. The box continues to move from the left to the right until it contacts the final curvature. The final curvatures are used for final closing of packing box at the top and the curvature will close the flaps at the box top gradually. There are two curvatures in final curvature mechanism: front curvature and rear curvature. The front curvature is applied to push down the short flaps and simultaneously separate

J. Zheng Li, *CAD, 3D Modeling, Engineering Analysis, and Prototype Experimentation*, DOI 10.1007/978-3-319-05921-1_9

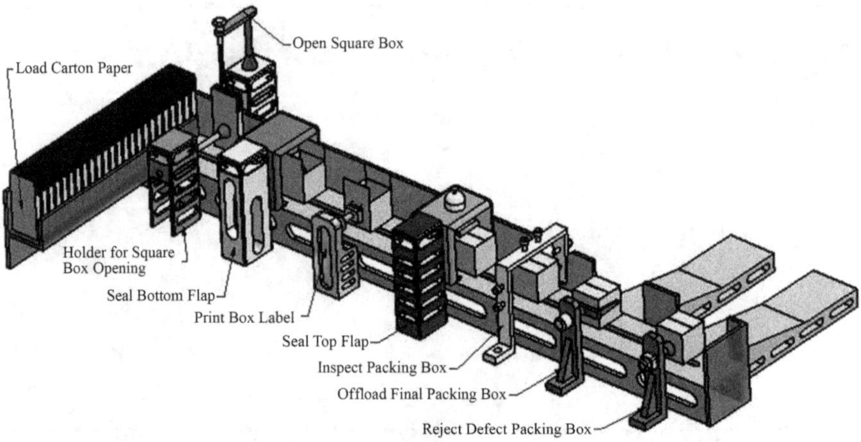

Fig. 9.1 Fully automated packaging assembly system

long flaps. This curvature is combined with two vertically and one horizontally changed curvatures to keep gradual closing of short flaps. The rear curvature is applied to gradually close two long flaps. The offloading mechanism is to load finished packing box to the product loading area and rejecting mechanism is to reject the packing box if some defects occurred during packing process.

9.2 Computer-Aided Simulation on Automated and High-Speed Packaging Machinery System

There are various loading forces generated during automated packaging processes including vibration, insertion, and packing. The computer-aided modeling and simulation are applied to diagnose the mechanism function and verify the component strength for solid performance. The vibration is caused by different forces including mass inertia and internal shocking forces during kinematic movement. The mathematical modeling shows the following equation (Isaev et al. 2005; Kundu and Cohen 2008):

$$M(F) = \frac{B_{\mathrm{BF}}(F)}{B_{\mathrm{OD}}(F)} = \frac{M''(F)}{M'(F)} \tag{9.1}$$

Here, $M'(F) = \frac{B_{\mathrm{OD}}(F)}{D_{\mathrm{P}}(F)}$ and $M''(F) = \frac{B_{\mathrm{BF}}(F)}{D_{\mathrm{P}}(F)}$.

$D_{\mathrm{P}}(F)$—Fourier transform of pushing device.

$B_{\mathrm{OD}}(F)$—Fourier transform of output delivery moving acceleration.

$B_{\mathrm{BF}}(F)$—Fourier transform of base frame acceleration.

Figures 9.2, 9.3, 9.4, 9.5, 9.6, 9.7, 9.8, 9.9, 9.10, 9.11, 9.12, 9.13, 9.14, 9.15, 9.16, 9.17, 9.18, 9.19, 9.20, 9.21, 9.22, 9.23, 9.24, 9.25, and 9.26 display the computational simulation results in this new automated and high-speed packaging system.

Fig. 9.2 Carton loading unit

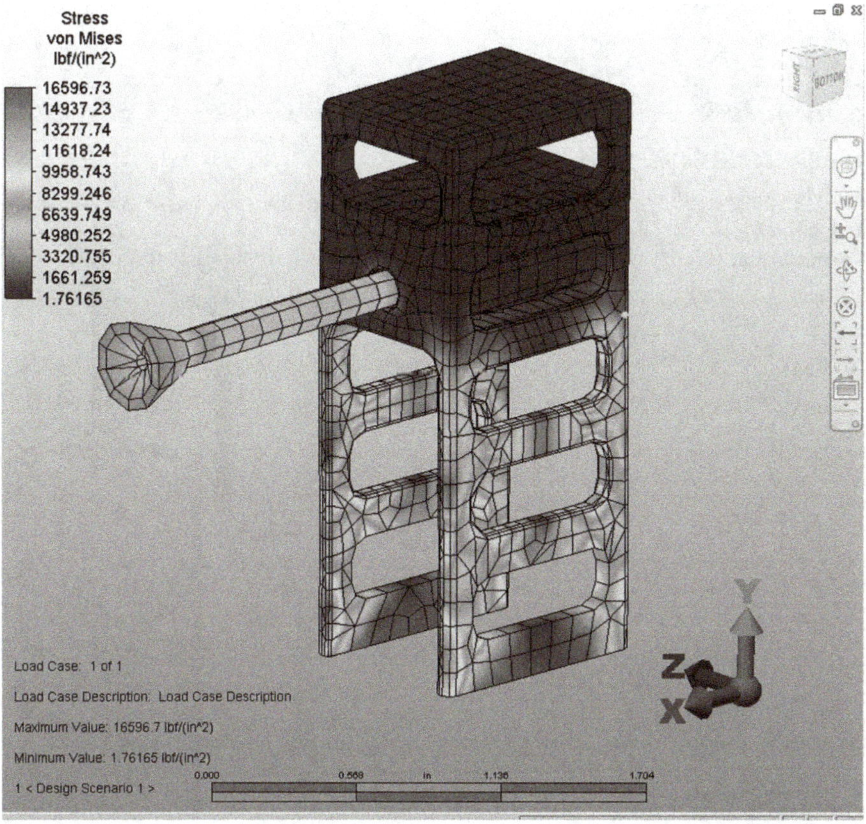

Fig. 9.3 Stress profile in carton loading unit

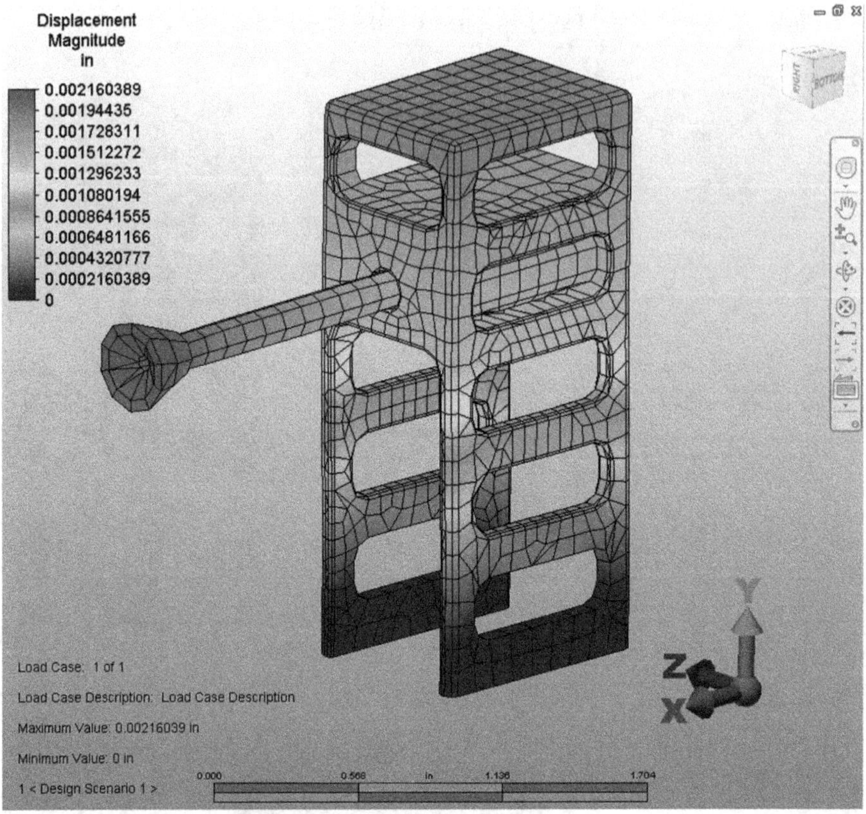

Fig. 9.4 Deflection profile in carton loading unit

Fig. 9.5 Carton separating unit

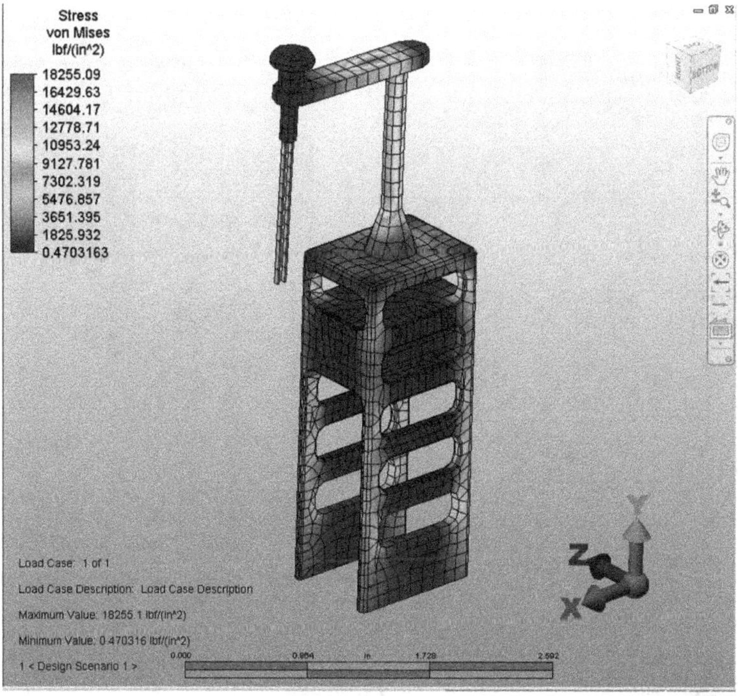

Fig. 9.6 Stress profile in carton separating unit

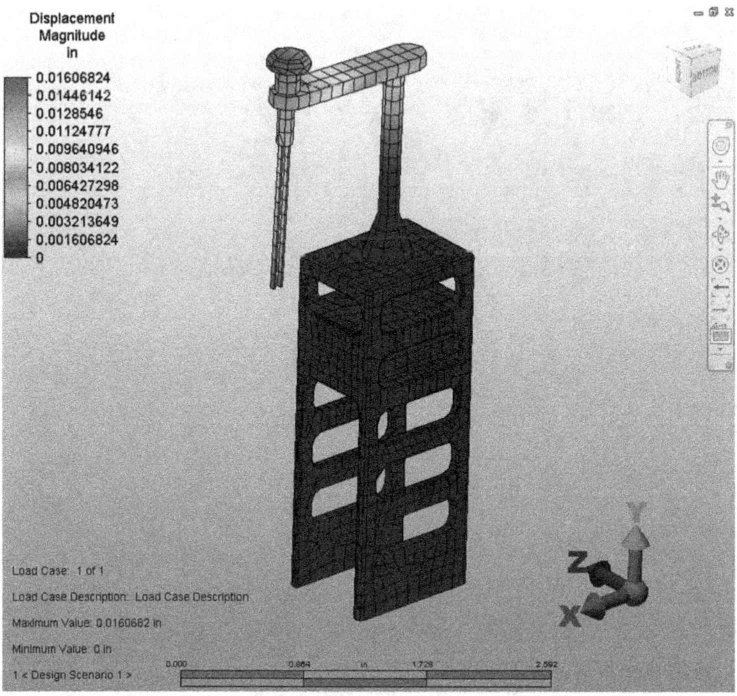

Fig. 9.7 Deflection profile in carton separating unit

Fig. 9.8 Carton bottom closer

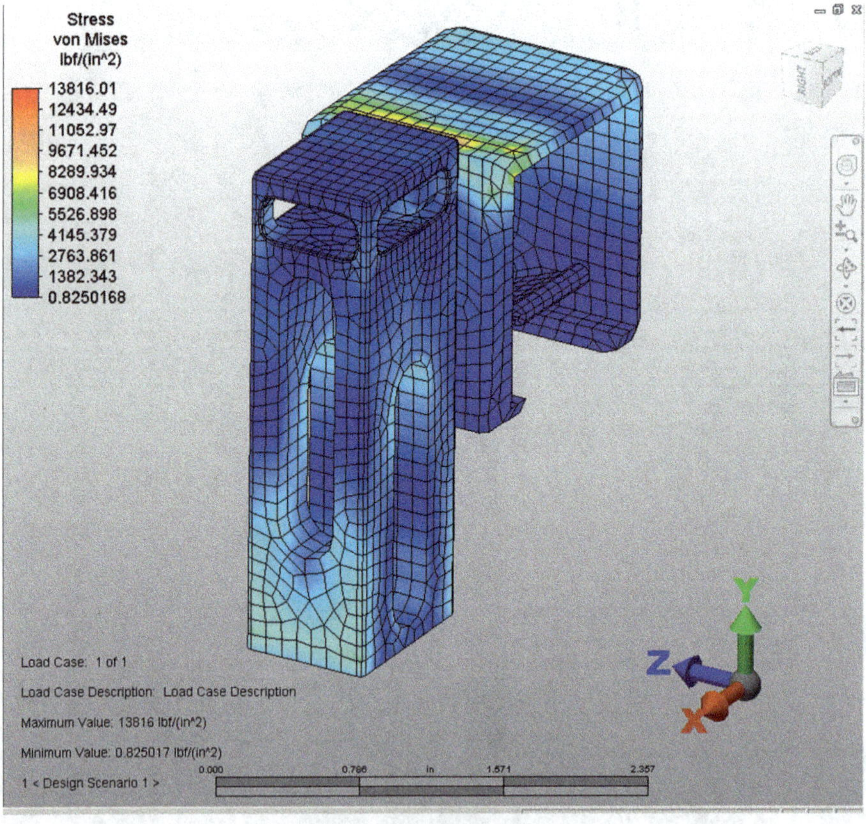

Fig. 9.9 Stress profile at base support in carton bottom closer

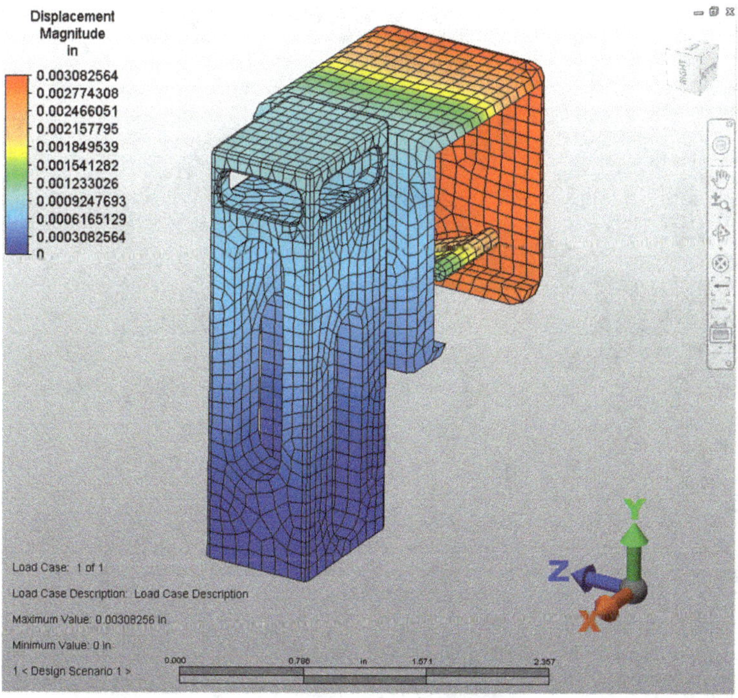

Fig. 9.10 Deflection profile at base support in carton bottom closer

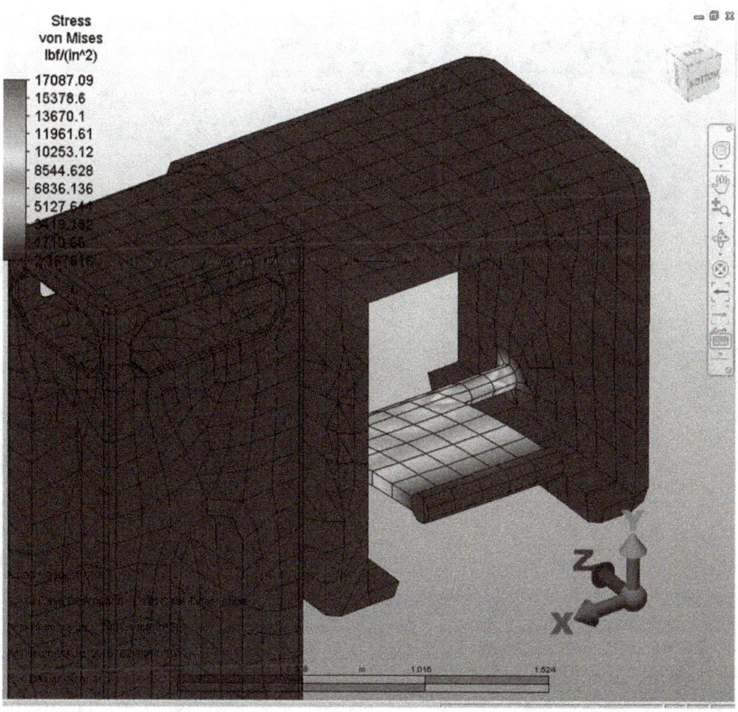

Fig. 9.11 Stress profile at tension mechanism in carton bottom closer

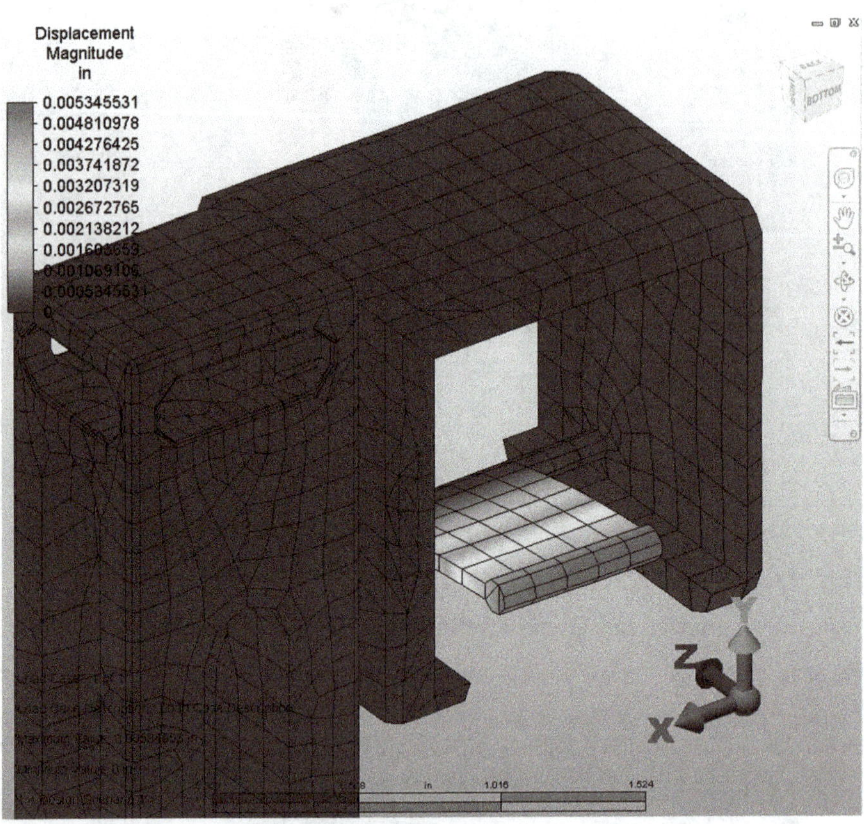

Fig. 9.12 Deflection profile at tension mechanism in carton bottom closer

Fig. 9.13 Labeling unit

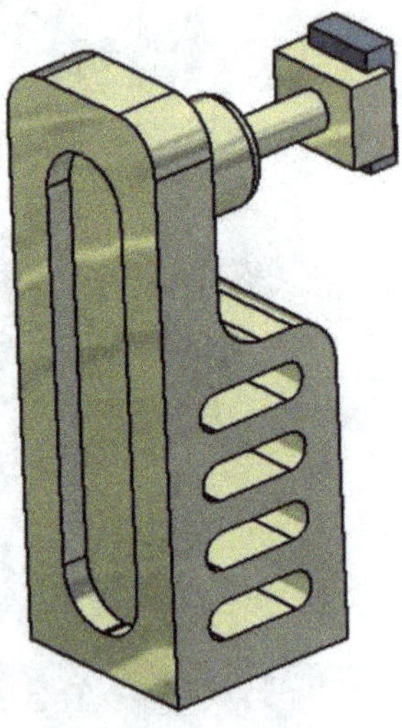

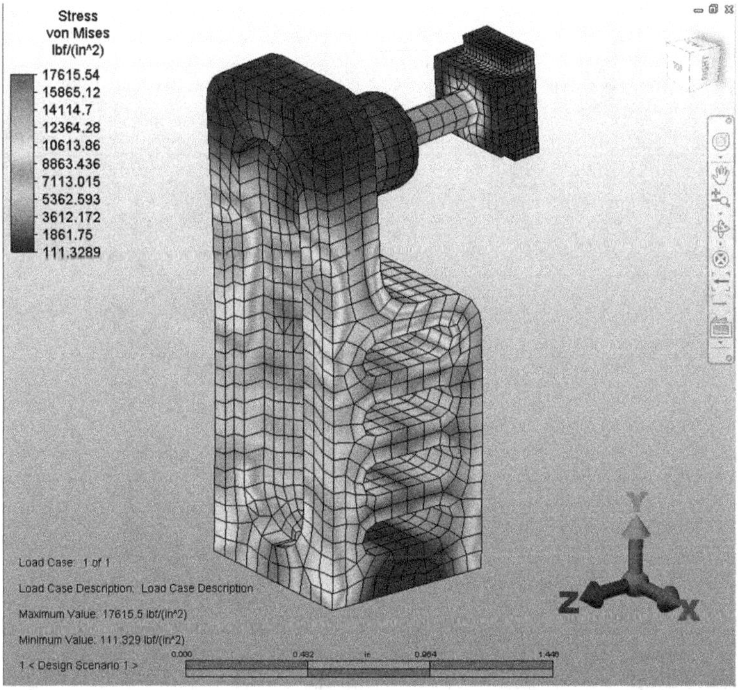

Fig. 9.14 Stress profile in labeling unit

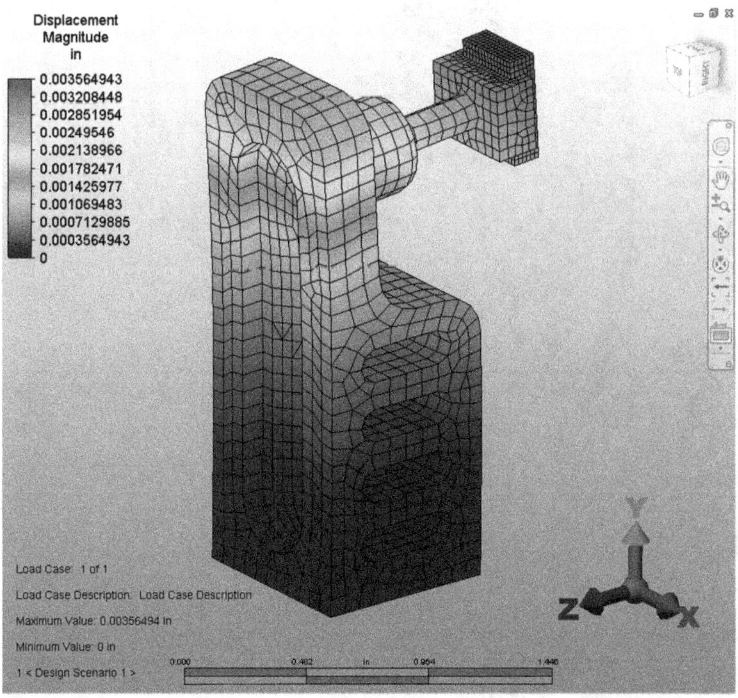

Fig. 9.15 Deflection profile in labeling unit

Fig. 9.16 Carton top closer

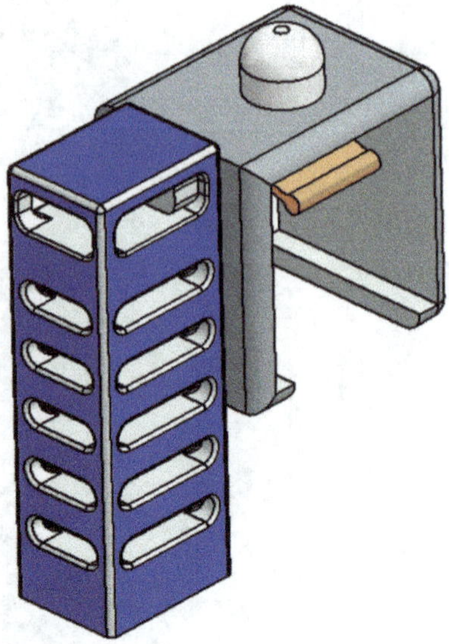

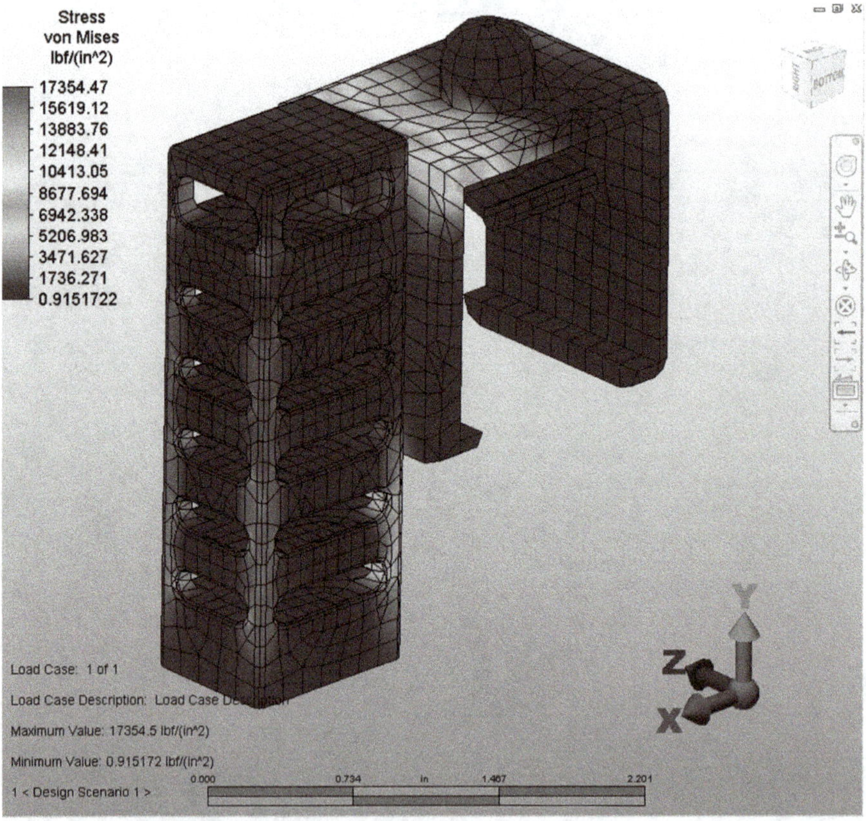

Fig. 9.17 Stress profile at base support in carton top closer

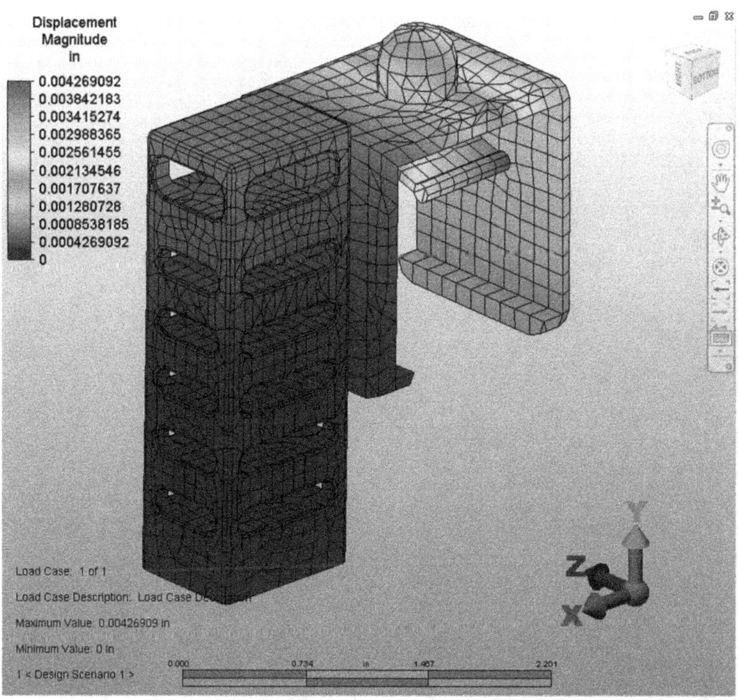

Fig. 9.18 Deflection profile at base support in carton top closer

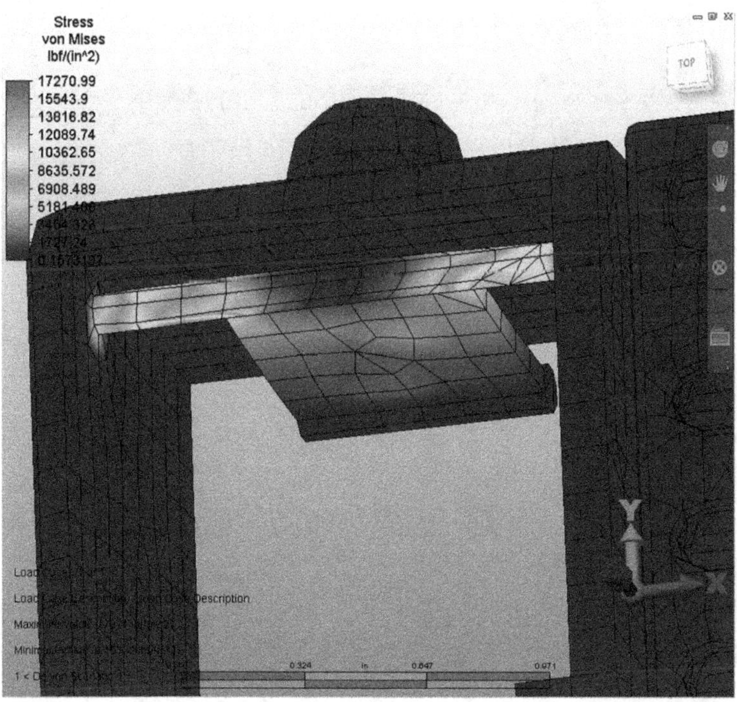

Fig. 9.19 Stress profile at tension mechanism in carton top closer

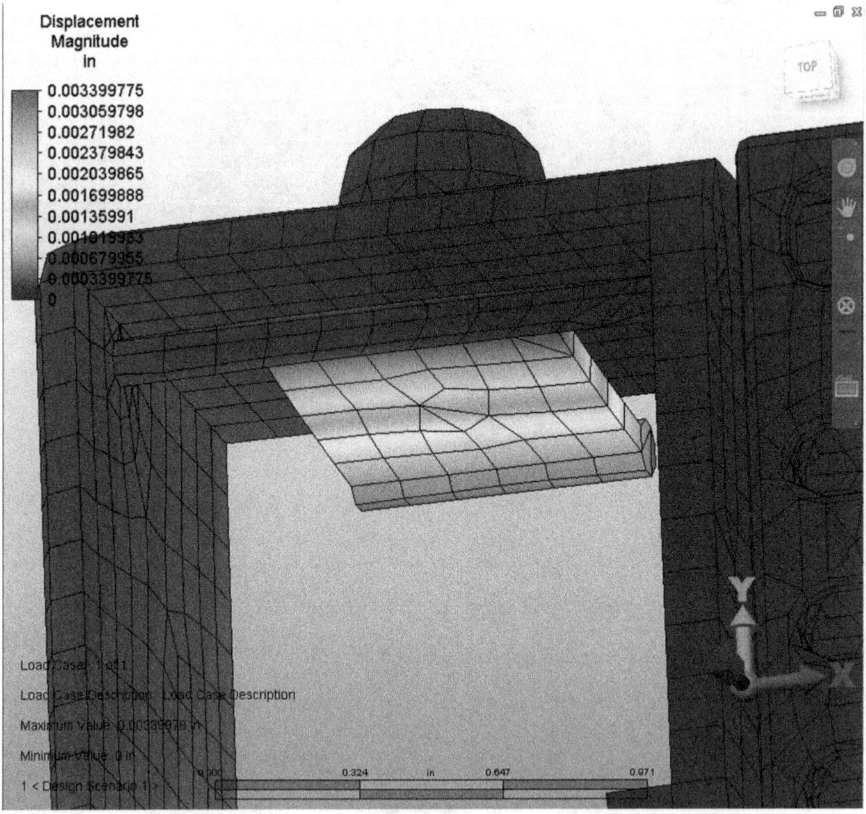

Fig. 9.20 Deflection profile at tension mechanism in carton top closer

Fig. 9.21 Offloading unit

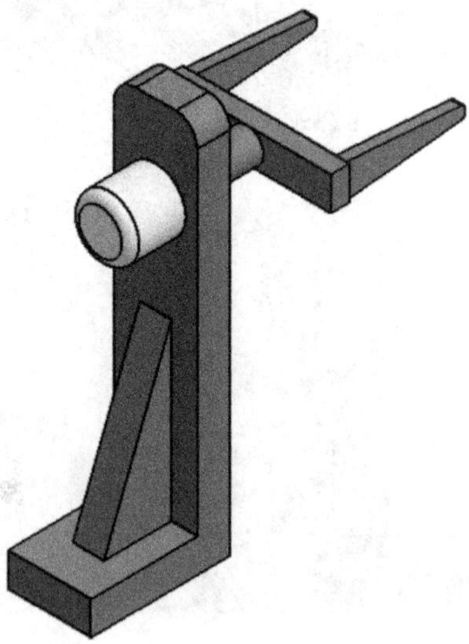

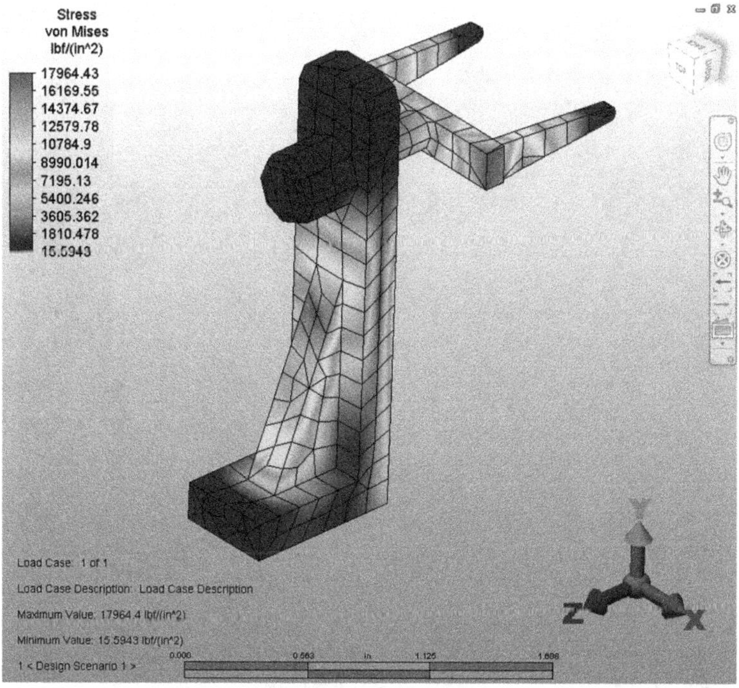

Fig. 9.22 Stress profile in offloading unit

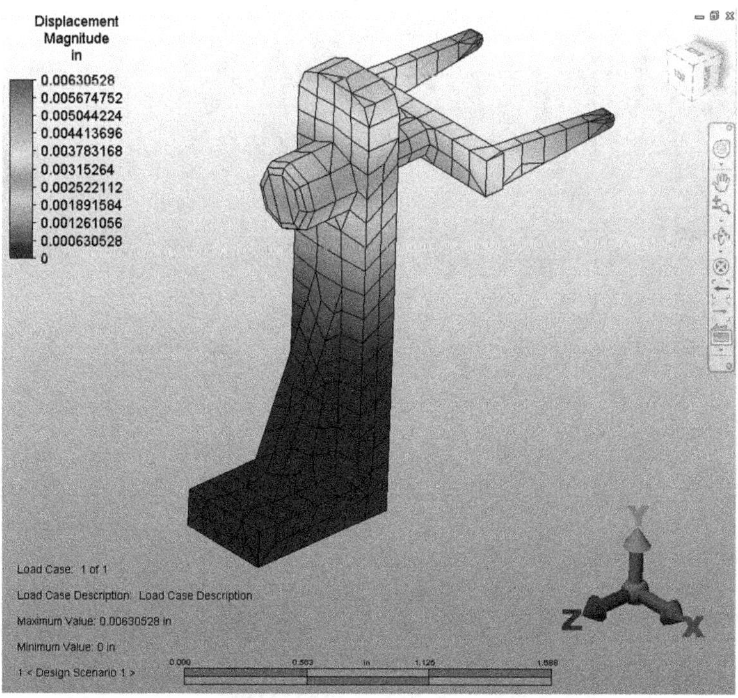

Fig. 9.23 Deflection profile in offloading unit

Fig. 9.24 Rejecting unit

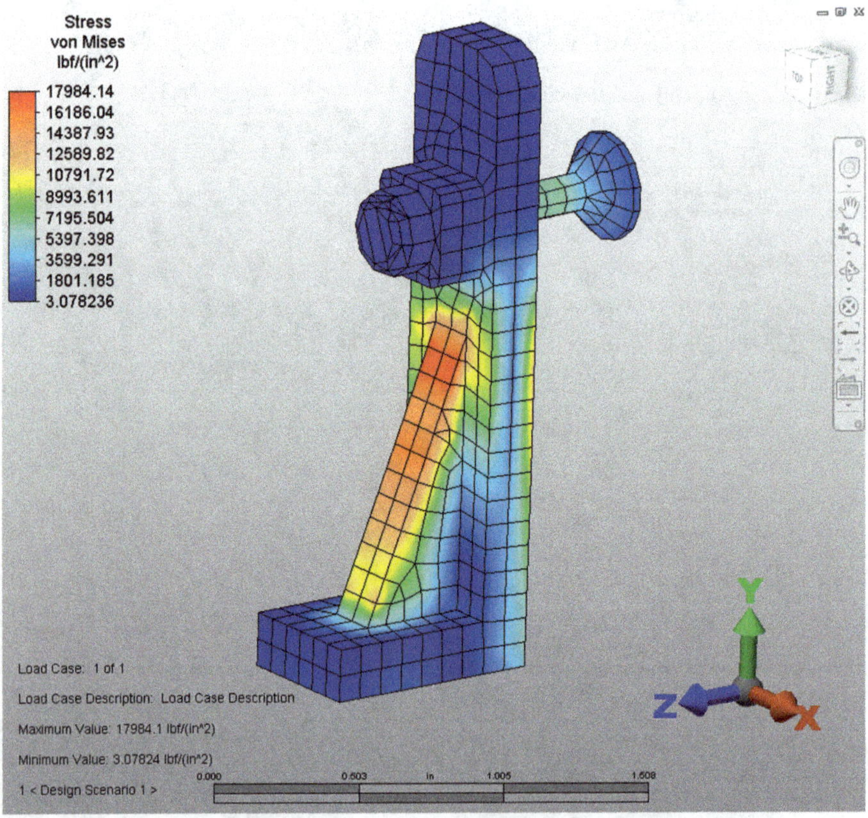

Fig. 9.25 Stress profile in rejecting unit

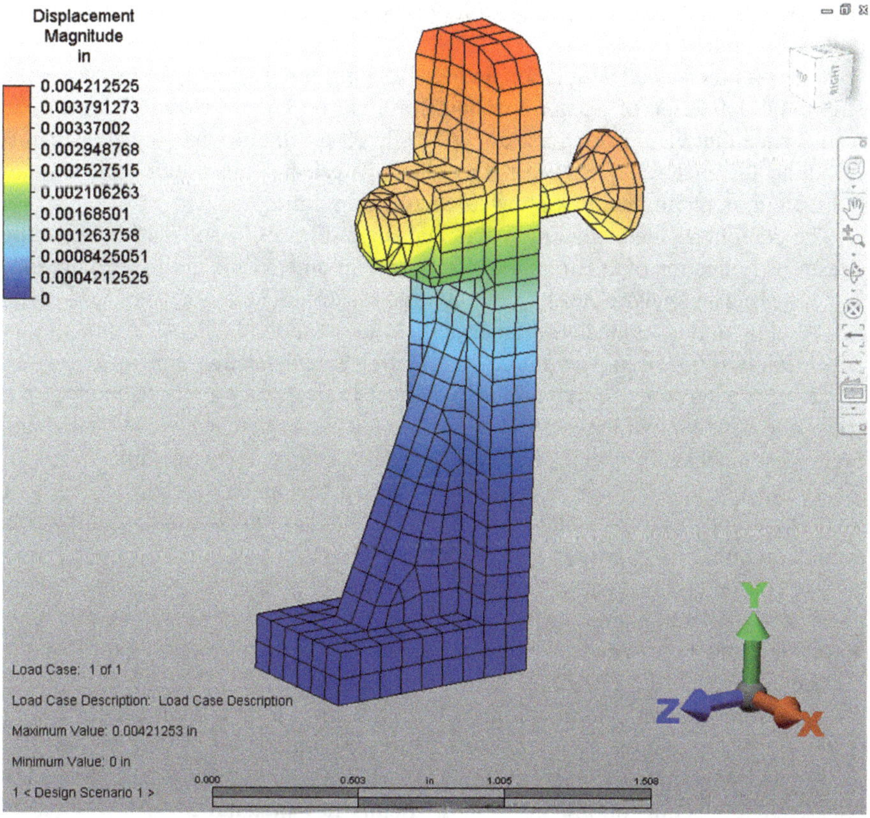

Fig. 9.26 Deflection profile in rejecting unit

The computer-aided simulation and analysis in Figs. 9.3 and 9.4 present the stress and deflection of carton loading unit in this new automated and high speed packaging system. The analytic results demonstrate that the maximum stress of 16,596.73 psi in this carton loading unit is less than the material yield strength of 36,000 psi and maximum deflection of 0.00216 in. is within material allowable deflection limit.

The computer-aided simulation and analysis in Figs. 9.6 and 9.7 indicate the stress and deflection of carton separating unit in this new automated and high-speed packaging system. The analytic results state that the maximum stress of 18,255.09 psi in this carton separating unit is less than the material yield strength of 36,000 psi and maximum deflection of 0.01607 in. is within material allowable deflection limit.

The computer-aided simulation and analysis in Figs. 9.9 and 9.10 indicate the stress and deflection of carton bottom closer in this new automated and high-speed packaging system. The analytic results state that the maximum stress of 13,816.01 psi in this carton bottom closer is less than the material yield strength of 36,000 psi and maximum deflection of 0.00308 in. is within material allowable deflection limit.

The computer-aided simulation and analysis in Figs. 9.11 and 9.12 indicate the stress and deflection of tension mechanism in carton bottom closer. The analytic results state that the maximum stress of 17,087.09 psi in this tension mechanism is less than the material yield strength of 36,000 psi and maximum deflection of 0.00535 in. is within material allowable deflection limit.

The computer-aided simulation and analysis in Figs. 9.14 and 9.15 indicate the stress and deflection of labeling unit in this new automated and high-speed packaging system. The analytic results state that the maximum stress of 17,615.54 psi in this labeling unit is less than the material yield strength of 36,000 psi and maximum deflection of 0.00356 in. is within material allowable deflection limit.

The computer-aided simulation and analysis in Figs. 9.17 and 9.18 indicate the stress and deflection of base support of carton top closer in this new automated and high-speed packaging system. The analytic results state that the maximum stress of 17,354.47 psi in this base support is less than the material yield strength of 36,000 psi and maximum deflection of 0.00427 in. is within material allowable deflection limit.

The computer-aided simulation and analysis in Figs. 9.19 and 9.20 indicate the stress and deflection of tension mechanism of carton top closer in this new automated and high-speed packaging system. The analytic results state that the maximum stress of 17,270.99 psi in this tension mechanism of carton top closer is less than the material yield strength of 36,000 psi and maximum deflection of 0.00339 in. is within material allowable deflection limit.

The computer-aided simulation and analysis in Figs. 9.22 and 9.23 indicate the stress and deflection of offloading unit in this new automated and high-speed packaging system. The analytic results state that the maximum stress of 17,964.43 psi in this offloading unit is less than the material yield strength of 36,000 psi and maximum deflection of 0.00631 in. is within material allowable deflection limit.

The computer-aided simulation and analysis in Figs. 9.25 and 9.26 indicate the stress and deflection of rejecting unit in this new automated and high-speed packaging system. The analytic results state that the maximum stress of 17,984.14 psi in this rejecting unit is less than the material yield strength of 36,000 psi and maximum deflection of 0.00421 in. is within material allowable deflection limit.

The above computational simulation results displayed in these figures show that the maximum stresses on these important components are all less than the material yield stress and maximum material deflections are all within material allowable deformation limits. The computational solutions confirm that this newly developed automated and high-speed packaging system works well in packaging applications.

9.3 Experiment on Automated and High-Speed Packaging Machinery System

The newly designed automated and high-speed packaging machinery system has been prototyped and tested to compare and verify the results from computer-aided simulation. Table 9.1 demonstrates the prototype testing results of carton loading unit in this new automated and high-speed packaging machinery system.

The prototype experimental results of carton loading unit in Table 9.1 verify the proper function because the average maximum stress 16,595.84 psi and average maximum deflection 0.00242 in. are close to the results of maximum stress 16,596.73 psi and maximum deflection 0.00216 in. that are represented, respectively, in Figs. 9.3 and 9.4 by computer-aided modeling and numerical simulation.

Table 9.2 expresses the prototype testing results of carton separating unit in this new automated and high-speed packaging machinery system.

The prototype experimental results of carton separating unit in Table 9.2 confirm the appropriate function since the average maximum stress 18,255.72 psi and average maximum deflection 0.01656 in. are similar to the results of maximum stress 18,255.09 psi and maximum deflection 0.01607 in. that are indicated, respectively, in Figs. 9.6 and 9.7 by computer-aided modeling and numerical simulation.

Table 9.3 records the prototype testing results of base support of carton bottom closer unit in this new automated and high-speed packaging machinery system.

Table 9.1 Prototype testing of carton loading unit in this new automated and high-speed packaging machinery system

Number of experiment	Maximum stress (psi)	Maximum deflection (in.)
1	16,595.78	0.00238
2	16,596.65	0.00256
3	16,594.38	0.00208
4	16,595.88	0.00255
5	16,596.98	0.00268
6	16,596.48	0.00212
7	16,596.55	0.00205
8	16,596.24	0.00286
9	16,594.95	0.00211
10	16,595.18	0.00202
11	16,596.35	0.00278
12	16,594.88	0.00266
13	16,595.18	0.00298
14	16,596.48	0.00206
15	16,594.65	0.00201
16	16,596.26	0.00256
17	16,596.59	0.00299
18	16,594.68	0.00245
19	16,596.22	0.00206
20	16,596.45	0.00236
Average	16,595.84	0.00242

Table 9.2 Prototype testing of carton separating unit in this new automated and high-speed packaging machinery system

Number of experiment	Maximum stress (psi)	Maximum deflection (in.)
1	18,256.38	0.01655
2	18,254.99	0.01588
3	18,255.59	0.01678
4	18,254.88	0.01599
5	18,255.18	0.01697
6	18,256.01	0.01678
7	18,255.09	0.01698
8	18,256.18	0.01638
9	18,255.48	0.01684
10	18,255.78	0.01678
11	18,256.35	0.01688
12	18,254.89	0.01598
13	18,255.85	0.01668
14	18,256.48	0.01695
15	18,255.88	0.01684
16	18,256.68	0.01597
17	18,254.99	0.01685
18	18,256.84	0.01678
19	18,254.98	0.01593
20	18,255.88	0.01648
Average	18,255.72	0.01656

Table 9.3 Prototype testing of base support of carton bottom closer unit in this new automated and high-speed packaging machinery system

Number of experiment	Maximum stress (psi)	Maximum deflection (in.)
1	13,816.01	0.00338
2	13,816.01	0.00304
3	13,815.38	0.00298
4	13,816.08	0.00368
5	13,815.18	0.00355
6	13,815.25	0.00299
7	13,816.11	0.00388
8	13,815.08	0.00398
9	13,816.15	0.00297
10	13,815.02	0.00378
11	13,815.18	0.00389
12	13,815.04	0.00395
13	13,816.15	0.00378
14	13,815.18	0.00296
15	13,815.06	0.00388
16	13,816.08	0.00393
17	13,815.15	0.00397
18	13,815.08	0.00298
19	13,815.12	0.00396
20	13,816.18	0.00388
Average	13,815.52	0.00357

Table 9.4 Prototype testing of tension mechanism of carton bottom closer unit in this new automated and high-speed packaging machinery system

Number of experiment	Maximum stress (psi)	Maximum deflection (in.)
1	17,087.02	0.00502
2	17,087.88	0.00555
3	17,088.89	0.00499
4	17,088.55	0.00508
5	17,086.88	0.00505
6	17,087.99	0.00575
7	17,086.98	0.00498
8	17,087.85	0.00438
9	17,087.98	0.00525
10	17,086.78	0.00478
11	17,087.68	0.00418
12	17,087.79	0.00556
13	17,086.78	0.00448
14	17,087.84	0.00418
15	17,086.97	0.00435
16	17,087.84	0.00448
17	17,087.98	0.00505
18	17,087.88	0.00435
19	17,086.78	0.00438
20	17,087.75	0.00538
Average	17,087.60	0.00486

The prototype experimental results for base support of carton bottom closer unit in Table 9.3 prove the normal function as the average maximum stress 13,815.52 psi and average maximum deflection 0.00357 in. are almost equal to the results of maximum stress 13,816.01 psi and maximum deflection 0.00308 in. that are laid out, respectively, in Figs. 9.9 and 9.10 by computer-aided modeling and numerical simulation.

Table 9.4 states the prototype testing results of tension mechanism of carton bottom closer unit in this new automated and high-speed packaging machinery system.

The prototype experimental results for tension mechanism of carton bottom closer unit in Table 9.4 verify the proper function because the average maximum stress 17,087.60 psi and average maximum deflection 0.00486 in. are very similar to the results of maximum stress 17,087.09 psi and maximum deflection 0.00535 in. that are shown, respectively, in Figs. 9.11 and 9.12 by computer-aided modeling and numerical simulation.

Table 9.5 demonstrates the prototype testing results of labeling unit in this new automated and high-speed packaging machinery system.

The prototype experimental results of labeling unit in Table 9.5 confirm the appropriate function since the average maximum stress 17,616.12 psi and average maximum deflection 0.00314 in. are closely equal to the results of maximum stress

Table 9.5 Prototype testing of labeling unit in this new automated and high-speed packaging machinery system

Number of experiment	Maximum stress (psi)	Maximum deflection (in.)
1	17,616.48	0.00318
2	17,615.38	0.00377
3	17,616.84	0.00308
4	17,616.78	0.00305
5	17,615.29	0.00268
6	17,616.98	0.00368
7	17,615.44	0.00278
8	17,616.68	0.00301
9	17,616.68	0.00367
10	17,616.74	0.00303
11	17,615.18	0.00302
12	17,615.98	0.00298
13	17,615.54	0.00388
14	17,616.48	0.00288
15	17,615.25	0.00248
16	17,616.48	0.00317
17	17,616.78	0.00338
18	17,615.38	0.00278
19	17,616.88	0.00248
20	17,615.18	0.00378
Average	17,616.12	0.00314

17,615.54 psi and maximum deflection 0.00356 in. that are presented, respectively, in Figs. 9.14 and 9.15 by computer-aided modeling and numerical simulation.

Table 9.6 demonstrates the prototype testing results of base support of carton top closer unit in this new automated and high-speed packaging machinery system.

The prototype experimental results for base support of carton top closer unit in Table 9.6 prove the normal function as the average maximum stress 17,354.02 psi and average maximum deflection 0.00468 in. are almost same as the results of maximum stress 17,354.47 psi and maximum deflection 0.00427 in. that are represented, respectively, in Figs. 9.17 and 9.18 by computer-aided modeling and numerical simulation.

Table 9.7 expresses the prototype testing results of tension mechanism of carton top closer unit in this new automated and high-speed packaging machinery system.

The prototype experimental results for tension mechanism of carton top closer unit in Table 9.7 verify the proper function because the average maximum stress 17,270.49 psi and average maximum deflection 0.00375 in. are approximately equal to the results of maximum stress 17,270.99 psi and maximum deflection 0.00340 in. that are indicated, respectively, in Figs. 9.19 and 9.20 by computer-aided modeling and numerical simulation.

Table 9.8 records the prototype testing results of offloading unit in this new automated and high-speed packaging machinery system.

Table 9.6 Prototype testing of base support of carton top closer unit in this new automated and high-speed packaging machinery system

Number of experiment	Maximum stress (psi)	Maximum deflection (in.)
1	17,354.08	0.00477
2	17,354.68	0.00408
3	17,353.75	0.00487
4	17,353.55	0.00498
5	17,354.54	0.00418
6	17,353.66	0.00499
7	17,353.48	0.00495
8	17,354.08	0.00498
9	17,354.11	0.00418
10	17,354.54	0.00496
11	17,353.57	0.00498
12	17,354.12	0.00408
13	17,354.88	0.00498
14	17,353.48	0.00492
15	17,353.38	0.00499
16	17,354.78	0.00419
17	17,353.55	0.00487
18	17,353.48	0.00489
19	17,354.18	0.00418
20	17,354.55	0.00448
Average	17,354.02	0.00468

Table 9.7 Prototype testing of tension mechanism of carton top closer unit in this new automated and high-speed packaging machinery system

Number of experiment	Maximum stress (psi)	Maximum deflection (in.)
1	17,270.89	0.00395
2	17,270.18	0.00388
3	17,271.08	0.00339
4	17,270.24	0.00398
5	17,270.08	0.00385
6	17,271.05	0.00396
7	17,270.19	0.00338
8	17,270.25	0.00336
9	17,270.38	0.00399
10	17,271.08	0.00397
11	17,270.18	0.00338
12	17,270.27	0.00399
13	17,270.08	0.00387
14	17,271.02	0.00398
15	17,270.05	0.00339
16	17,270.12	0.00338
17	17,270.08	0.00398
18	17,271.05	0.00394
19	17,270.48	0.00398
20	17,271.05	0.00337
Average	17,270.49	0.00375

Table 9.8 Prototype testing of offloading unit in this new automated and high-speed packaging machinery system

Number of experiment	Maximum stress (psi)	Maximum deflection (in.)
1	17,964.25	0.00624
2	17,964.55	0.00638
3	17,963.48	0.00618
4	17,964.18	0.00584
5	17,963.38	0.00602
6	17,964.12	0.00608
7	17,963.24	0.00579
8	17,964.54	0.00601
9	17,963.21	0.00635
10	17,964.48	0.00575
11	17,964.24	0.00584
12	17,963.18	0.00603
13	17,964.42	0.00608
14	17,964.11	0.00579
15	17,963.23	0.00601
16	17,964.78	0.00603
17	17,963.11	0.00566
18	17,964.68	0.00575
19	17,963.48	0.00602
20	17,964.54	0.00638
Average	17,963.96	0.00601

The prototype experimental results of offloading unit in Table 9.8 confirm the appropriate function since the average maximum stress 17,963.96 psi and average maximum deflection 0.00601 in. are approximately same as the results of maximum stress 17,964.43 psi and maximum deflection 0.00631 in. that are laid out, respectively, in Figs. 9.22 and 9.23 by computer-aided modeling and numerical simulation.

Table 9.9 states the prototype testing results of rejecting unit in this new automated and high-speed packaging machinery system.

The prototype experimental results of rejecting unit in Table 9.9 prove the normal function as the average maximum stress 17,984.53 psi and average maximum deflection 0.00455 in. are close to the results of maximum stress 17,984.14 psi and maximum deflection 0.00421 in. that are shown, respectively, in Figs. 9.25 and 9.26 by computer-aided modeling and numerical simulation.

Table 9.9 Prototype testing of rejecting unit in this new automated and high-speed packaging machinery system

Number of experiment	Maximum stress (psi)	Maximum deflection (in.)
1	17,984.24	0.00438
2	17,984.84	0.00418
3	17,984.02	0.00455
4	17,984.18	0.00477
5	17,984.88	0.00495
6	17,984.95	0.00411
7	17,984.98	0.00488
8	17,984.09	0.00459
9	17,983.88	0.00411
10	17,984.48	0.00484
11	17,984.77	0.00488
12	17,984.05	0.00418
13	17,984.78	0.00415
14	17,984.95	0.00479
15	17,984.88	0.00499
16	17,984.08	0.00448
17	17,984.68	0.00415
18	17,984.97	0.00484
19	17,984.78	0.00498
20	17,984.08	0.00419
Average	17,984.53	0.00455

9.4 Discussion and Future Improvement on Automated and High-Speed Packaging Machinery Systems

Automated packaging machines are the multi-axis mechanical systems that require multiple interacting mechanisms to manipulate the components for final product packages. It can help industry to minimize unnecessary human involvement, reduce labor cost, improve packing quality, and increase productivity. The technology improvement helps developing and making the automatic packing machine simpler and more precised for accurate and flexible automatic packing operations. Future improvement will focus on modifying this new system for quick changeover on different product packing, easier tool and fixture adjustment, continuous workflow maintenance, consistent production line balance, and simple equipment installation.

Part III

Biomedical Systems

Biomedical and Surgical Systems 10

10.1 Design of Biomedical and Surgical Systems

In this new open surgical instrument design, the surgiclip delivery mechanism is improved if compared with conventional surgical instruments. The surgiclips are moving to the front jaw pair through distal move of surgiclip pusher which can be manipulated by instrument handles. When doctors and surgeons move instrument handles together, the surgiclip will be driven forward and completely formed after surgiclip pusher distally drives surgiclip into the guiding groove in the front jaw pair. When doctors and surgeons free the instrument handles, the surgiclip pusher automatically moves back to original location or starting position to catch next surgiclip for next firing cycle. Since the advancing process of this surgiclip can be easily and reliably manipulated in the new design, the highly required dimensional tolerance control will be not necessary in making the instrumental components during production process. This improvement can prevent the surgiclips from drop-off, simplify the machining process, raise the production rate, and reduce the manufacturing cost (Figs. 10.1 and 10.2).

This biomedical instrument is first positioned onto patient body tissues and then surgiclips are distally moved into the guiding track of front jaw pair driven by surgiclip pusher and the patient tissues securely clamped as doctors and surgeons manipulate to close the instrument handles. As soon as doctors and surgeons free the instrument handles, the front jaw pair will be open and surgiclip pusher and driving mechanism move back to its initial locations. Compared with the traditional surgiclip delivering mechanism, in which the surgiclips are moving forward into the guiding track of jaw pair by compression spring that sometime accidentally drop off the surgiclip from jaw pair track in the instrument, the surgiclip movement to the guiding track of jaw pair in this new instrument design can be smoothly guided and reliably controlled by surgeons. The driving mechanism that is connected to the pivot point of instrument handles distally travels to move forward the surgiclip into

J. Zheng Li, *CAD, 3D Modeling, Engineering Analysis, and Prototype Experimentation*, DOI 10.1007/978-3-319-05921-1_10

Fig. 10.1 New open surgiclip biomedical instrument

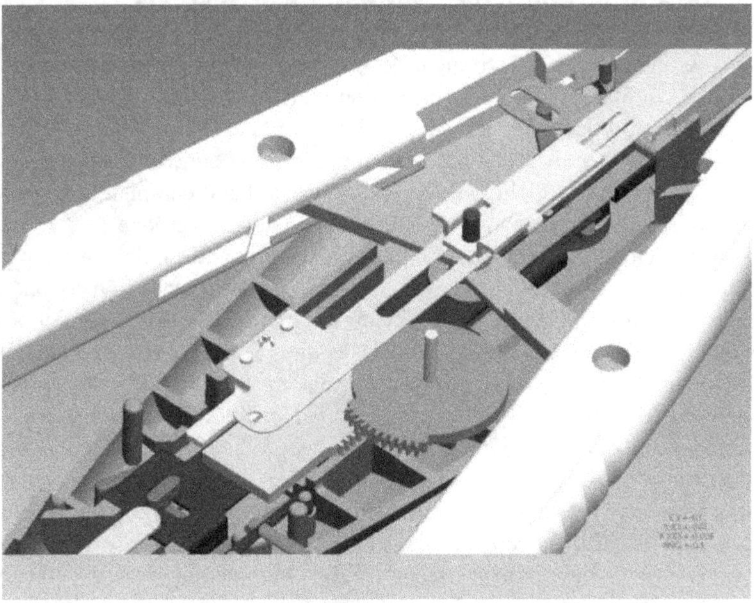

Fig. 10.2 Internal structure of new open surgiclip biomedical instrument

the guiding track of jaw pair and finally fully formed the surgiclip as doctors and surgeons gradually bring the instrument handles together. Such surgiclip linear movement can be smoothly and properly manipulated by surgeons to prevent the surgiclips from accidental drop-off from instrumental guiding track. The repeated prototype testing has demonstrated its stable and reliable performances of this new biomedical instrument design with no surgiclip drop-off and hand force to operate this new instrument is lower than the conventional biomedical surgiclip instrument.

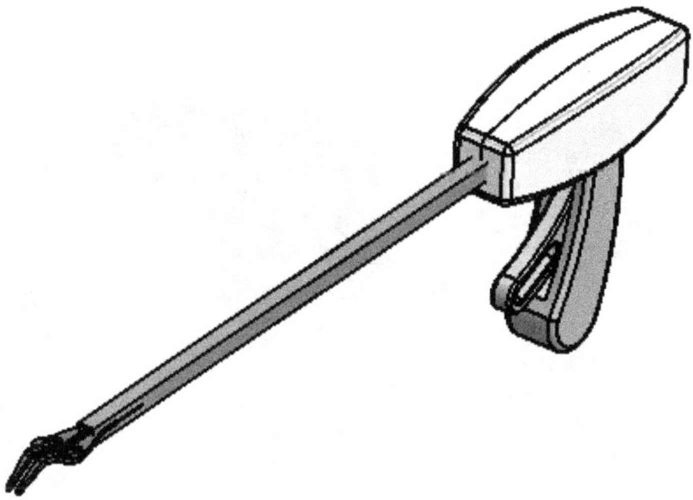

Fig. 10.3 Endo surgiclip biomedical instrument—view 1

Fig. 10.4 Endo surgiclip biomedical instrument—view 2

The newly designed endo surgiclip biomedical instrument is shown in Figs. 10.3, 10.4, and 10.5. It can deliver multiple surgiclips into front jaw guiding channel. The surgiclips which sit in lower position are guided to move vertically inside curved channel by moving block that is driven by surgiclip pusher. Since there must be no gap between surgiclips, the moving block has to be pushed to continuously contact the last surgiclip to keep distal movement of moving block only one-way direction.

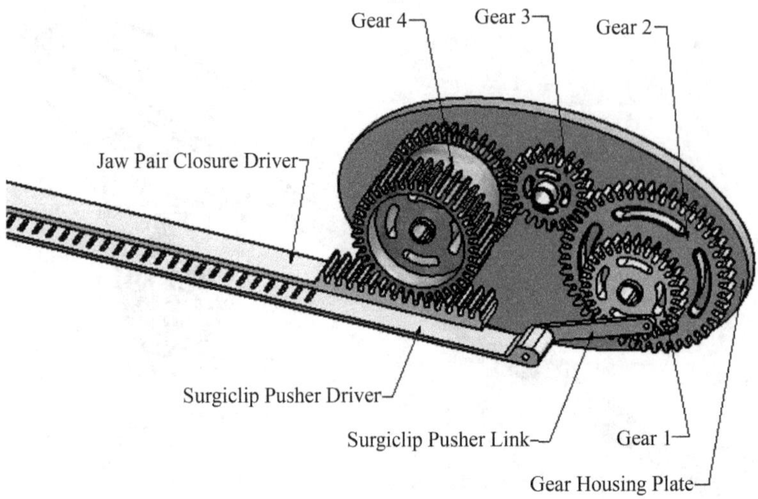

Fig. 10.5 Internal driving mechanism of endo surgiclip biomedical instrument

The moving block consists of the block holder and two snaps which are driven by hocking top and bottom compression springs. The top snap contacts the fixed gear rack and bottom snap contacts the movable gear rack which can travel up and down vertically.

At the start position, the movable gear rack moves from the right to the left. The front tip of the bottom snap engages with the teeth in movable gear rack and top snap slides over the teeth of movable gear rack. At step one, there is no relative movement between moving block and movable gear rack but there is movement between block holder and moving block. At step two, moving block distally moves and movable gear rack stops. When movable gear rack moves back (from the left to the right), the top snap engages the teeth in block holder and bottom snap slides distally over the teeth on movable gear rack. At step three, there is no movement between moving block and block holder but there is movement between movable gear rack and moving block. At step four, the movable gear rack moves to its original starting position and both snaps engage the teeth in movable gear rack. During cyclic movement from step one to step four, the movable block distally travels one clip distance along the surgiclip track.

The new design of this driving mechanism ensures one-way movement of moving block along the distal motion of surgiclips. Since length of each surgiclip is 0.256 in., the moving block needs to travel 0.276 in. to properly engage each surgiclip in this new driving mechanism design. Each time when surgiclips distally move one clip distance, the most front surgiclip of inline surgiclips will be pushed into the guiding channel in instrument jaw pair for further formation. The surgiclip pusher driven by driving mechanism moves distally to deliver surgiclip and travels proximally to pick up next surgiclip. The front part of surgiclip pusher delivers surgiclip into jaw pair and rear part links to the driving mechanism. In this new design, the surgiclip pusher requires to distally move 0.98 in. from its original position to deliver each surgiclip into final location of inside guiding channel in jaw pair and move proximally back to its original position to pick up net surgiclip. The driving mechanism supplies both 0.98 in. pusher movement and forces required to overcome the frictional forces and weights of all components including moveable gear rack, moving block, block holder, and surgiclip pusher.

10.2 Computer-Aided Simulation on Biomedical and Surgical Systems

The computer-aided simulations on biomedical open surgiclip instrument and biomedical endoscopic surgiclip instrument are described as follows.

10.2.1 Biomedical Open Surgiclip Instrument

The mathematic equation of force and energy balance in this new biomedical open surgiclip instrument design can be demonstrated as follows (Kundu and Cohen 2008):

$$P_{\text{load}} \times G_{\text{linear}} = Q \times \omega \tag{10.1}$$

Since the torque $Q = P_{\text{pivot}} \times L_{\text{R}}$,

$$Q \times \omega = P_{\text{pivot}} \times L_{\text{R}} \times \omega = P_{\text{pivot}} \times G_{\text{angular}} \tag{10.2}$$

Therefore,

$$P_{\text{load}} \times G_{\text{linear}} = P_{\text{pivot}} \times G_{\text{angular}} \tag{10.3}$$

$$P_{\text{load}} = \left(G_{\text{angular}}/G_{\text{linear}}\right) \times P_{\text{pivot}} = (GR) \times P_{\text{pivot}} \tag{10.4}$$

Referring the geometry design of this new instrument:

$$P_{\text{finger}} \times W = P_{\text{pivot}} \times L_{\text{R}} \tag{10.5}$$

The different operation force P_{finger} can be determined by combining different L_{R} and W values. The computer-aided simulation can be applied to find out the optimal combined values of L_{R} and W. The computer-aided solution suggested that the optimized performance of this new biomedical instrument will be achieved when $L_{\text{R}} = 4.78$ in. and $W = 2.18$ in. are selected.

From Eq. (7.3),

$$P_{\text{finger}} \times 4.78 = P_{\text{pivot}} \times 2.18$$

Fig. 10.6 Driving gear

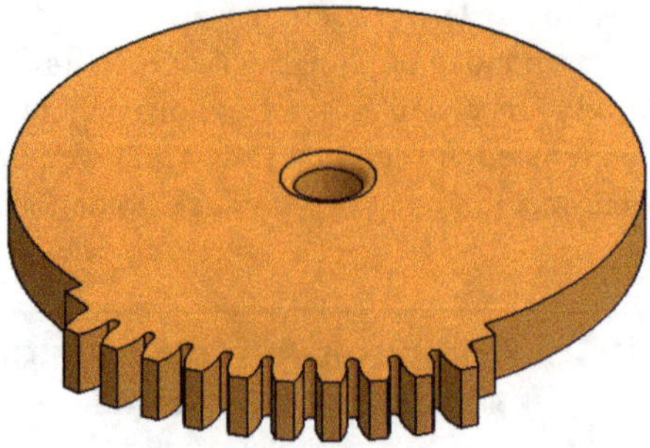

$$P_{\text{pivot}} = 2.192 \times P_{\text{finger}}$$

From Eq. (7.2),

$$P_{\text{load}} = (GR) \times P_{\text{pivot}} = (GR) \times 2.192 \times P_{\text{finger}}$$

The ratio of angular speed to linear speed ($G_{\text{angular}}/G_{\text{linear}}$) can be solved by computer-aided simulation to determine the optimal instrument performance. The one mechanical advantage of this new biomedical instrument can be determined when instrument handle has been fully closed:

$$\text{Mechanical advantage} = (GR) \times 2.192 \tag{10.6}$$

$$= (0.04838/0.03538) \times 2.192 = 2.997$$

This result indicates that when 20 lbf forces are required to fully form the biomedical surgiclip, the surgeons need to put 3.34 lbf forces to fully close the instrument handle which is less than the normal spec of 4 lbf in surgical operation procedure. This will ease doctors and surgeons in their medical treatments and surgical operations.

Figures 10.6, 10.7, 10.8, 10.9, 10.10, 10.11, 10.12, 10.13, 10.14, 10.15, 10.16, 10.17, 10.18, 10.19, 10.20, 10.21, 10.22, 10.23, 10.24, 10.25, 10.26, 10.27, 10.28, 10.29, 10.30, and 10.31 display the stress and deflection profiles of critical components in this new biomedical open surgiclip instrument.

The computer-aided simulation and analysis in Figs. 10.7 and 10.8 indicate the stress and deflection of driving gear in this new biomedical open surgical instrument. The analytic results state that the maximum stress of 23,663.32 psi in this driving gear is less than the material yield strength of 36,000 psi and maximum deflection of 0.00077 in. is within material allowable deflection limit.

The computer-aided simulation and analysis in Figs. 10.10 and 10.11 display the stress and deflection of trigger in this new biomedical open surgical instrument. The analytic results state that the maximum stress of 1,380.27 psi in this trigger is less than the material yield strength and maximum deflection of 0.00077 in. is within material allowable deflection limit.

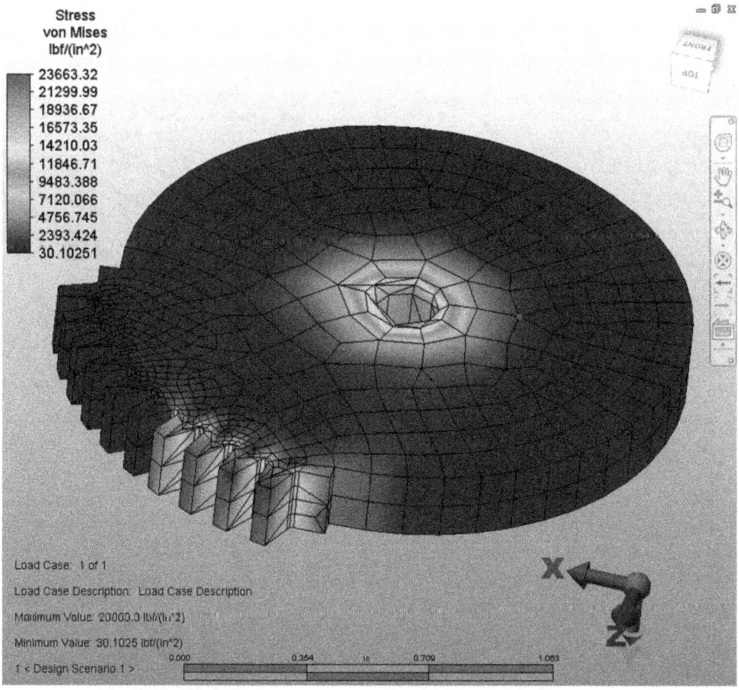

Fig. 10.7 Stress profile of driving gear

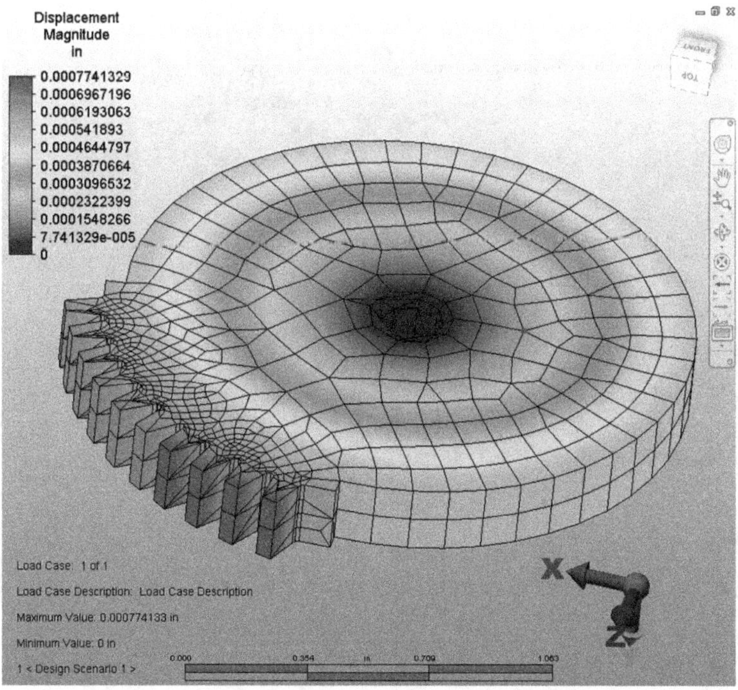

Fig. 10.8 Deflection profile of driving gear

Fig. 10.9 Trigger

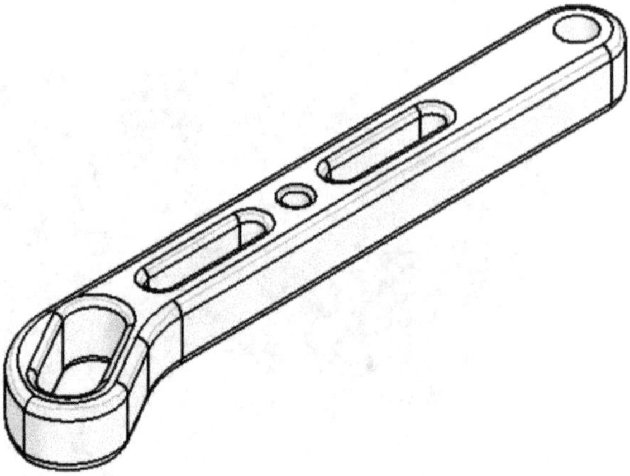

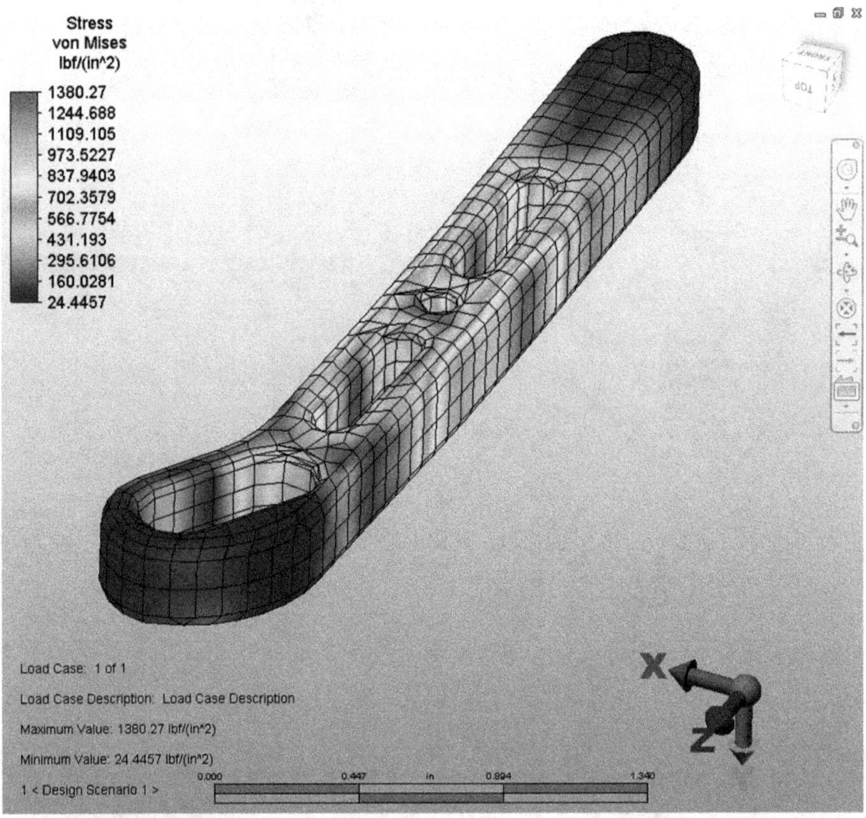

Fig. 10.10 Stress profile of trigger

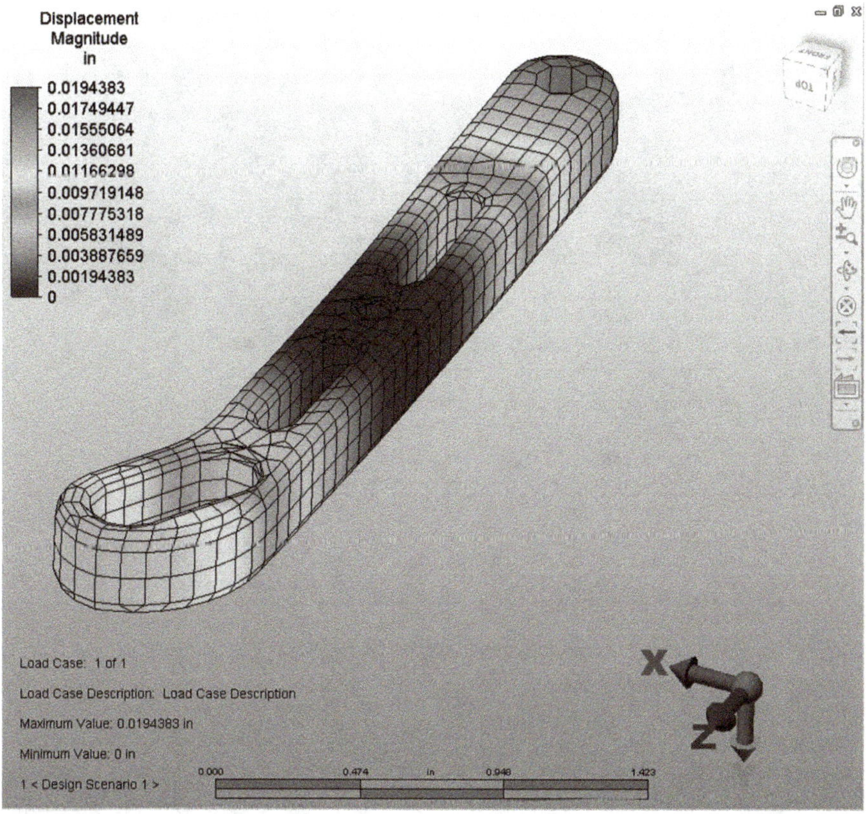

Fig. 10.11 Deflection profile of trigger

Fig. 10.12 Instrument front end of internal shaft

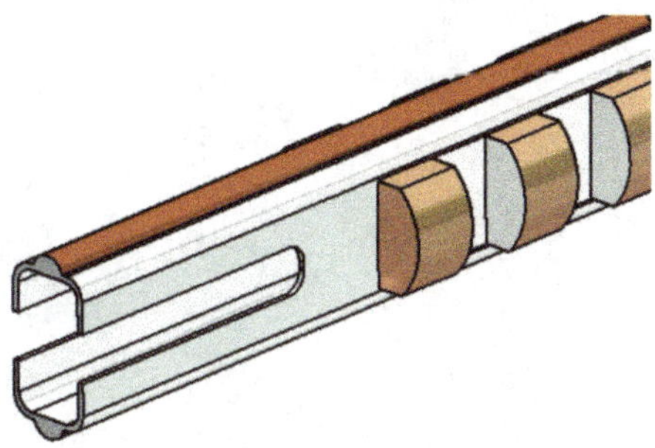

The following diagram shows the stress profile of internal driving shaft when surgeon manipulates patient body tissues during surgical procedures.

The computer-aided simulation and analysis in Figs. 10.13 and 10.14 show the stress and deflection of front end of internal shaft in this new biomedical open surgical instrument. The analytic results display that the maximum stress of

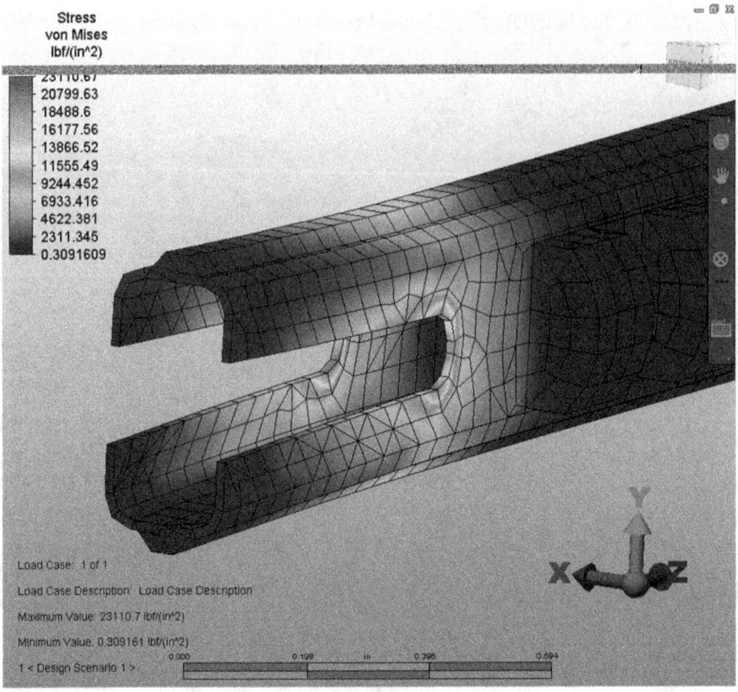

Fig. 10.13 Stress profile of instrument front end of internal shaft

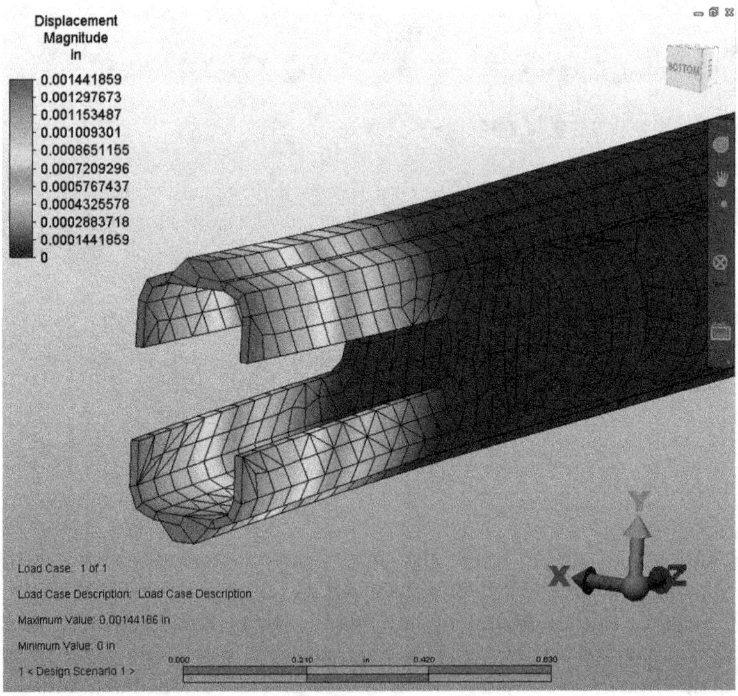

Fig. 10.14 Deflection profile of instrument front end of internal shaft

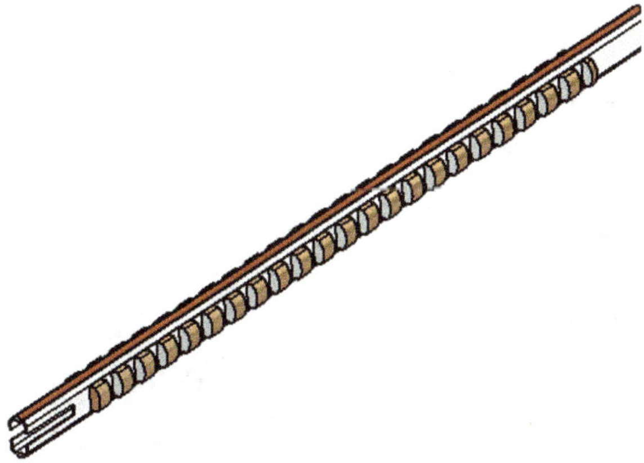

Fig. 10.15 Instrument internal shaft

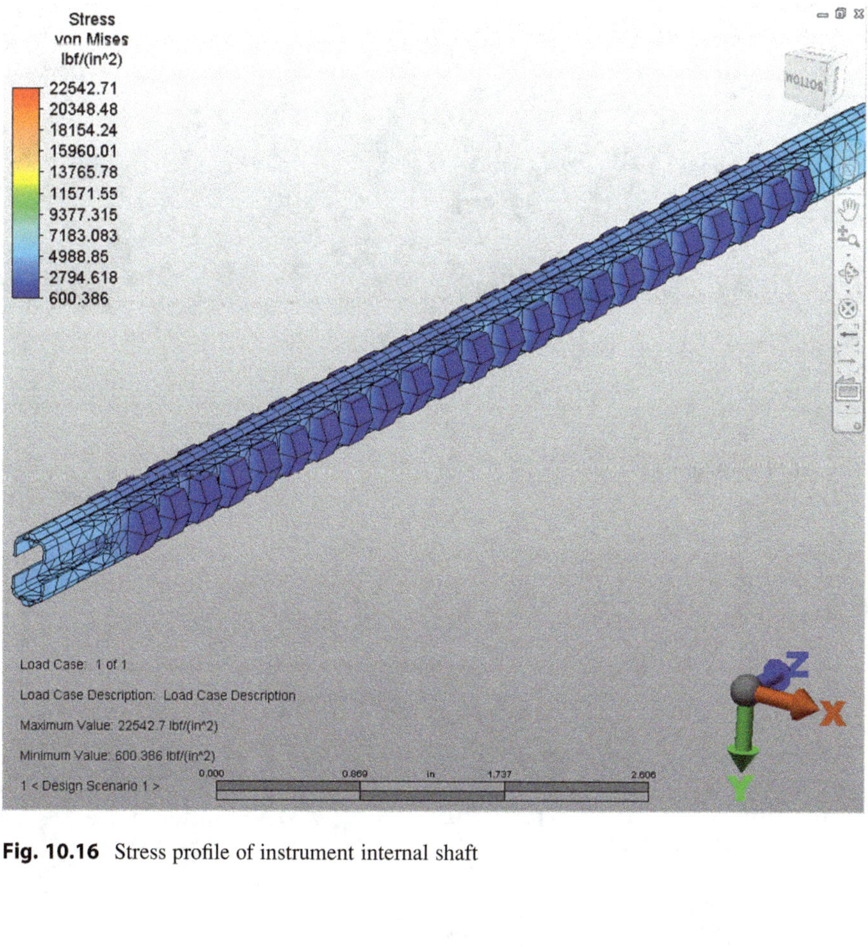

Fig. 10.16 Stress profile of instrument internal shaft

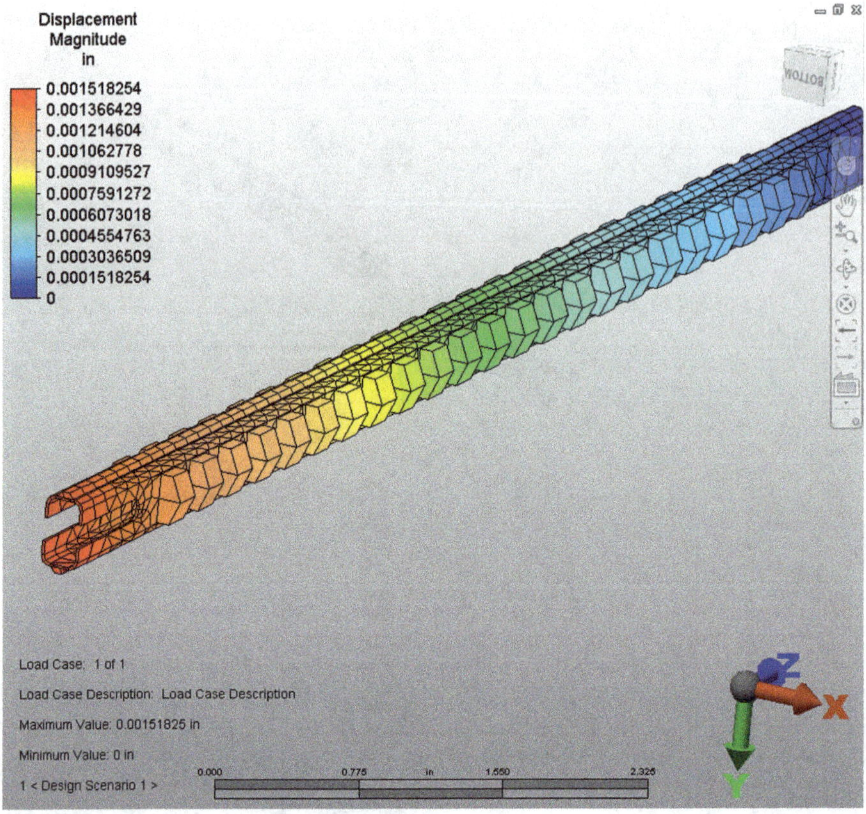

Fig. 10.17 Deflection profile of instrument internal shaft

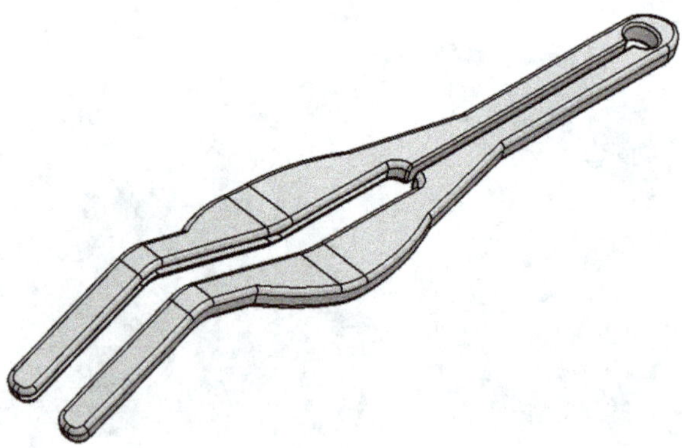

Fig. 10.18 Instrument jaw in closing tissue

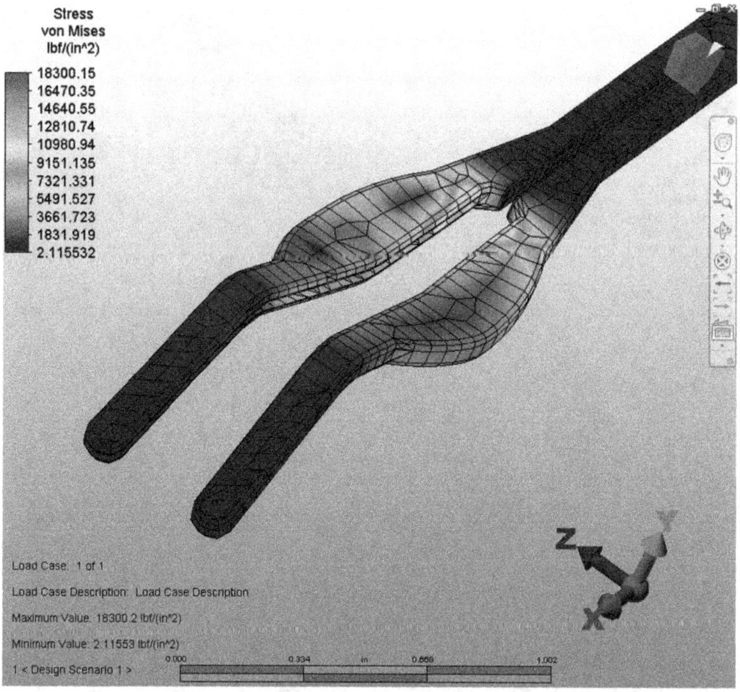

Fig. 10.19 Stress profile of instrument jaw in closing tissue

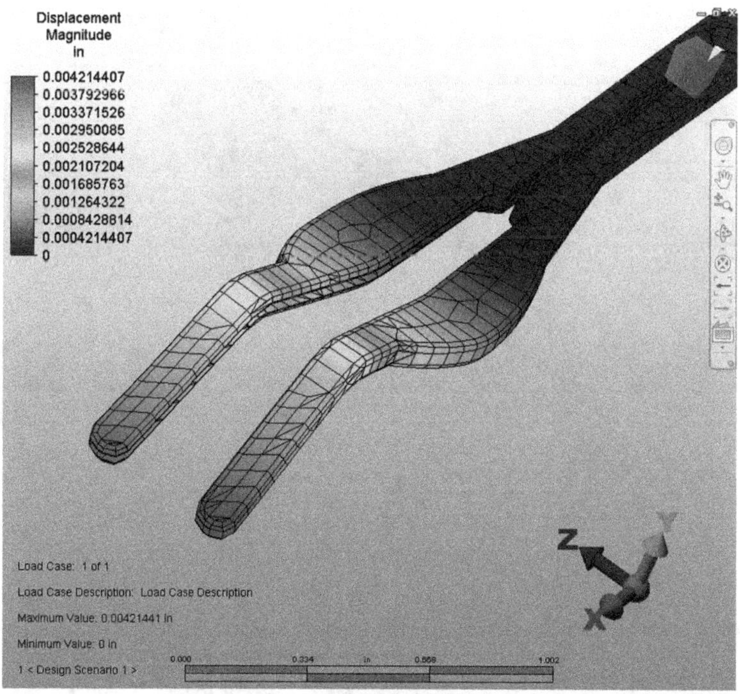

Fig. 10.20 Deflection profile of instrument jaw in closing tissue

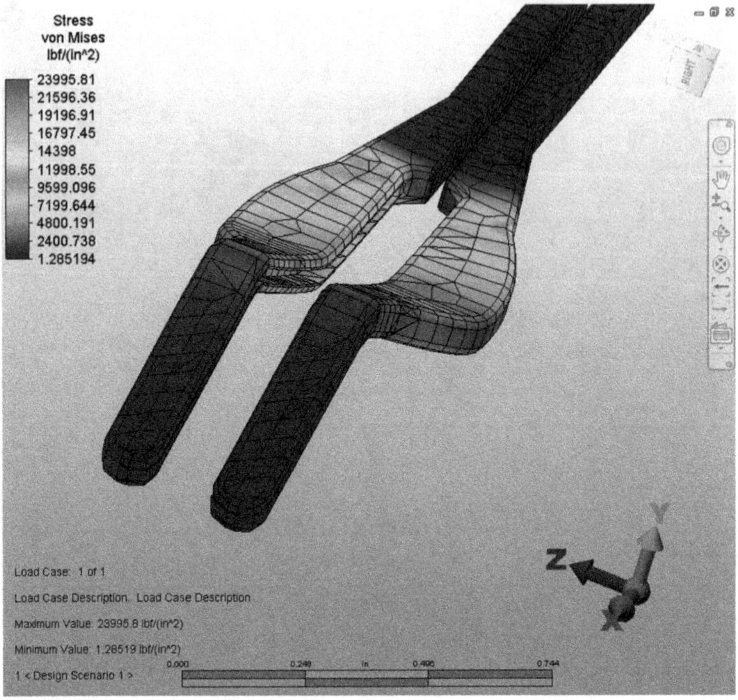

Fig. 10.21 Stress profile of instrument jaw in thick tissue manipulation

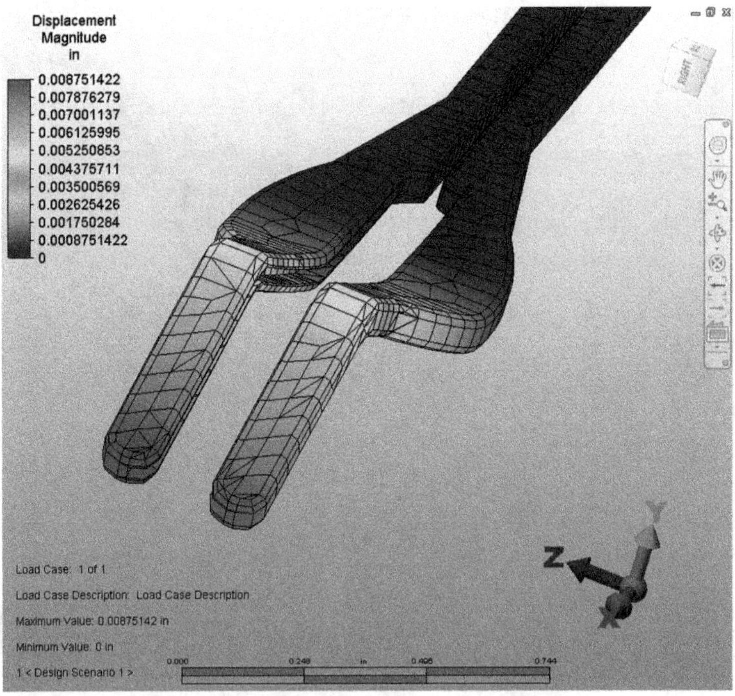

Fig. 10.22 Deflection profile of instrument jaw in thick tissue manipulation

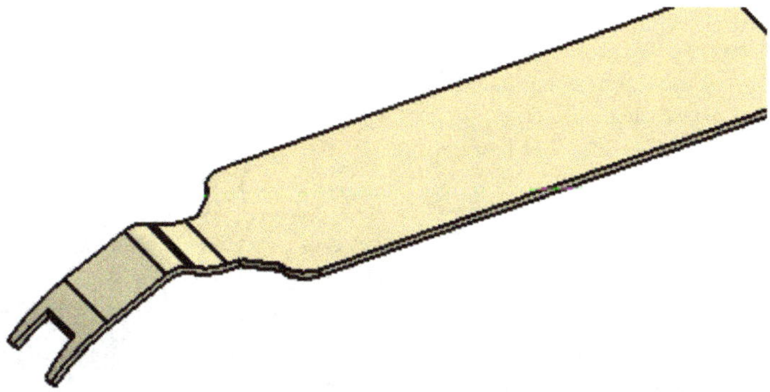

Fig. 10.23 Instrument surgiclip pusher

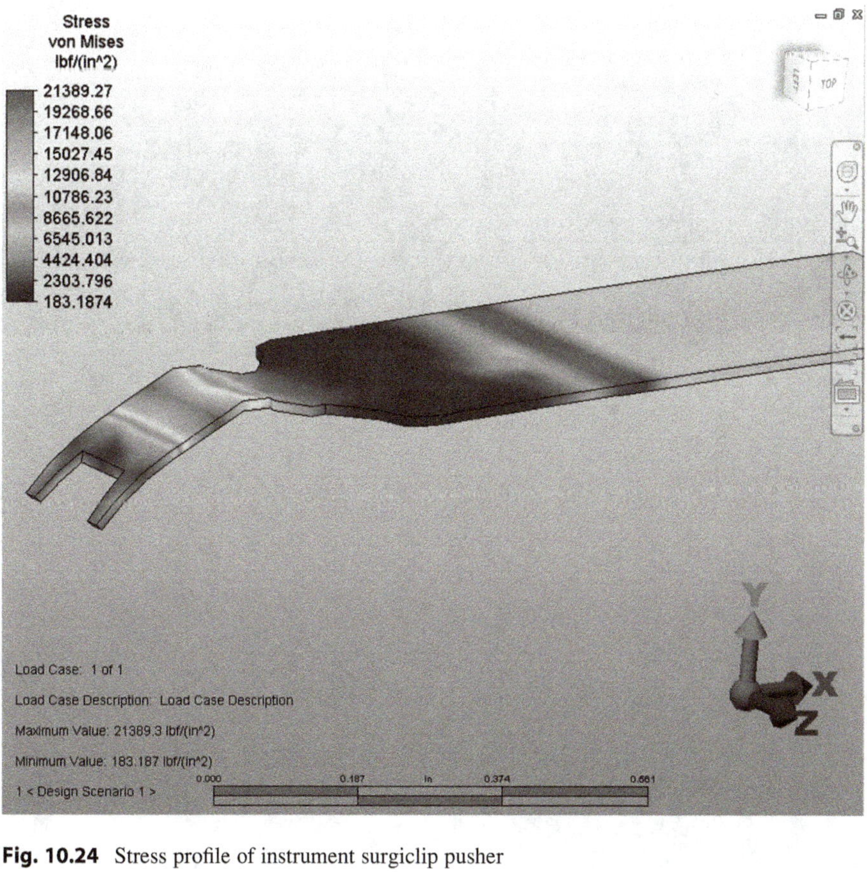

Fig. 10.24 Stress profile of instrument surgiclip pusher

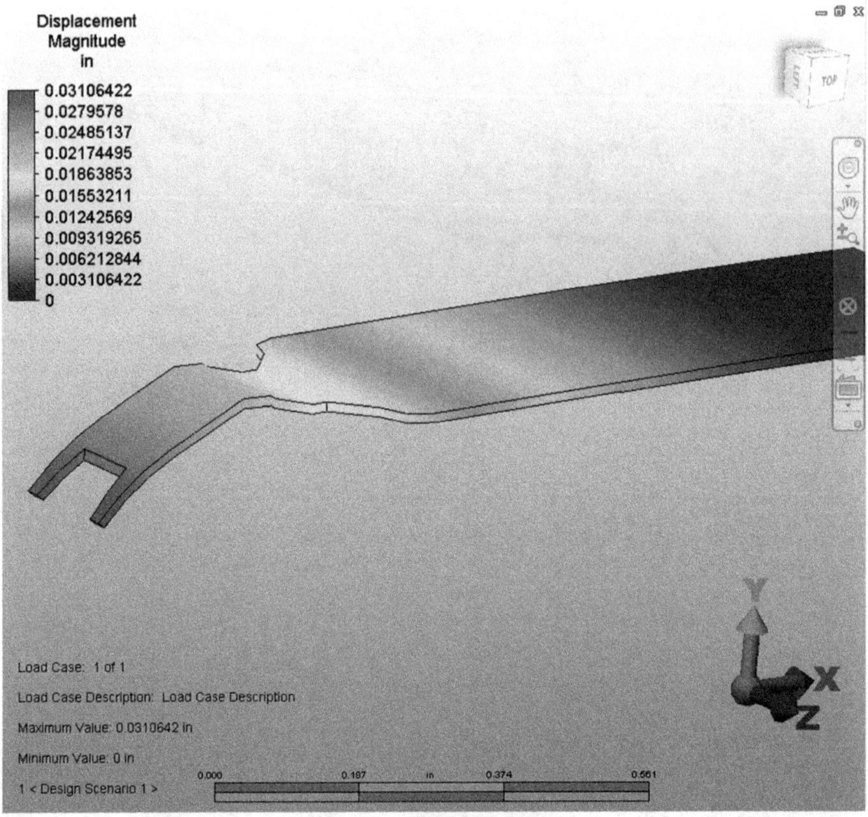

Fig. 10.25 Deflection profile of instrument surgiclip pusher

Fig. 10.26 Instrument external shaft

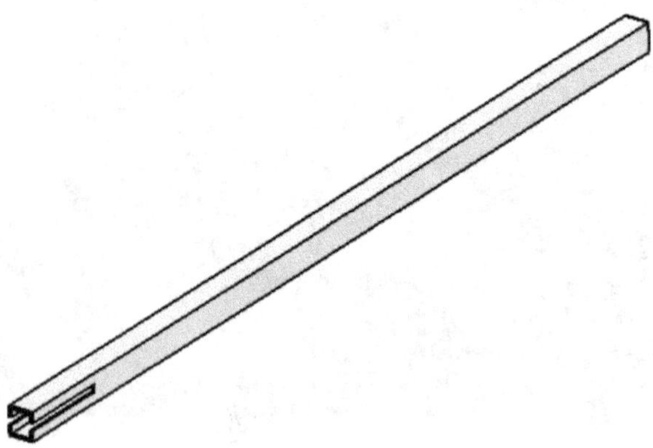

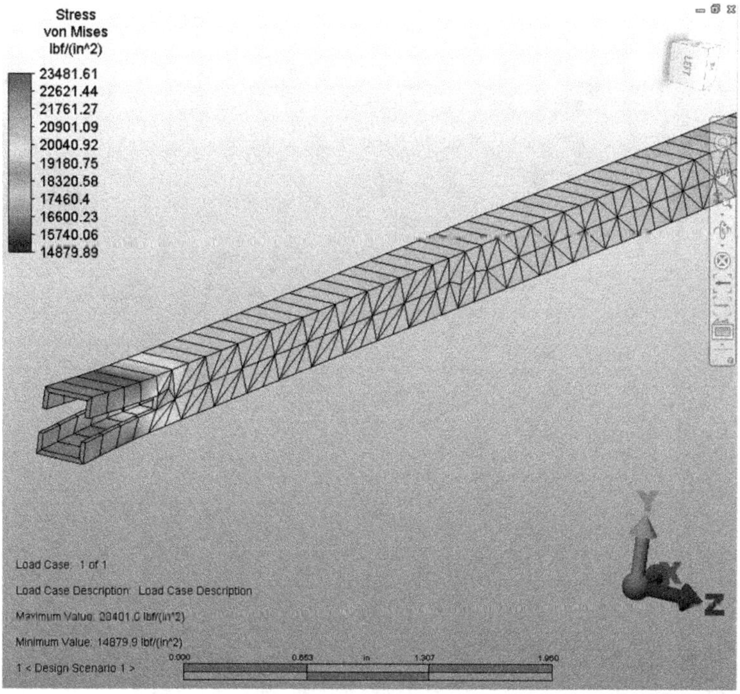

Fig. 10.27 Stress profile of instrument external shaft

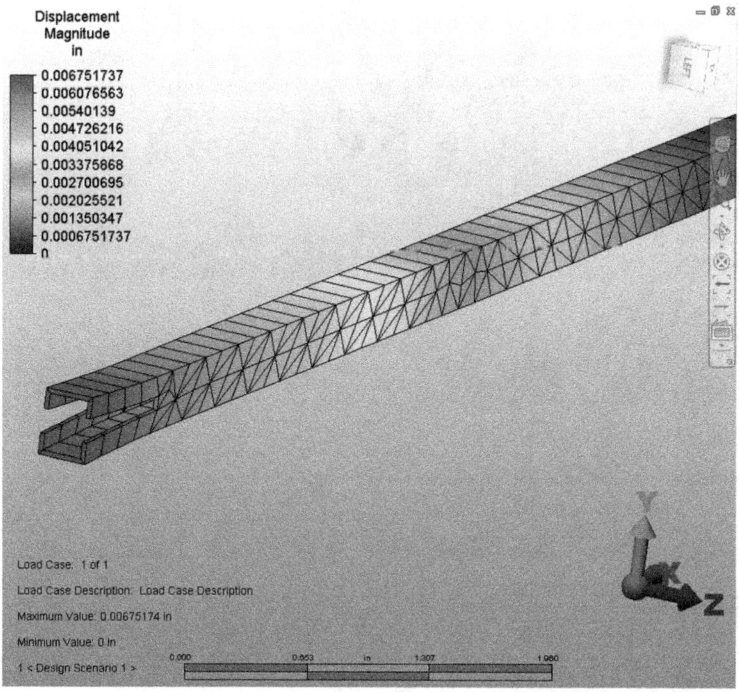

Fig. 10.28 Deflection profile of instrument external shaft

Fig. 10.29 Driving link

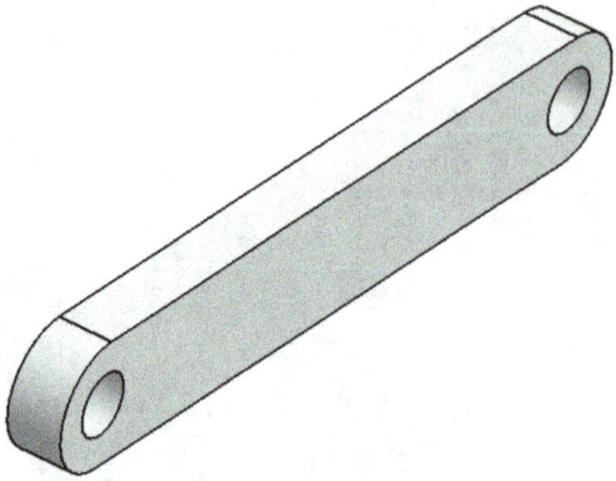

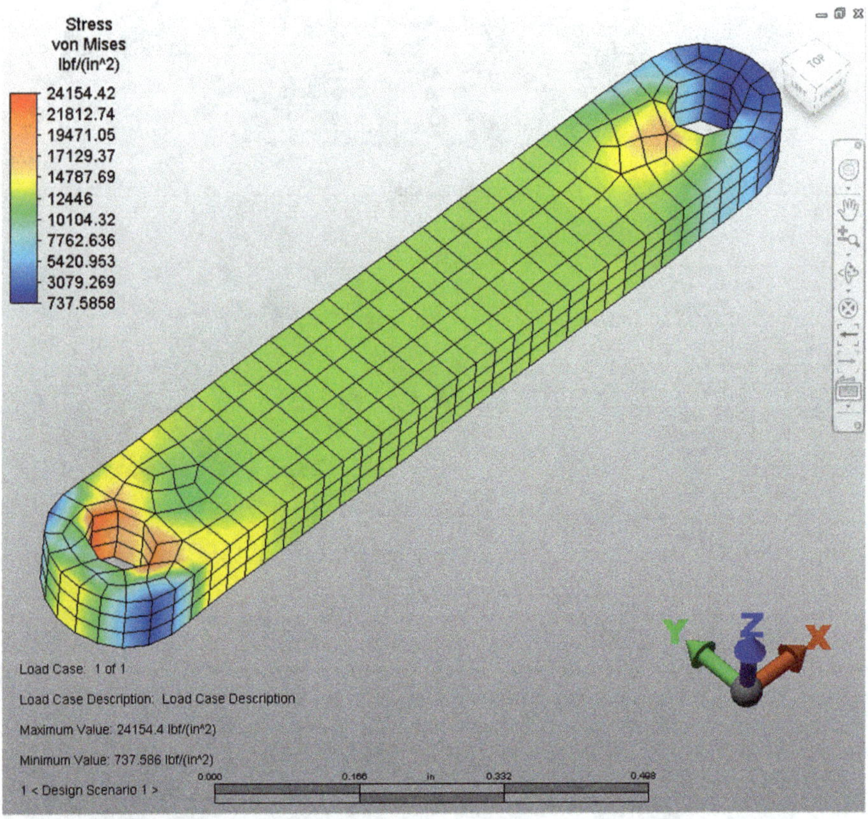

Fig. 10.30 Stress profile of driving link

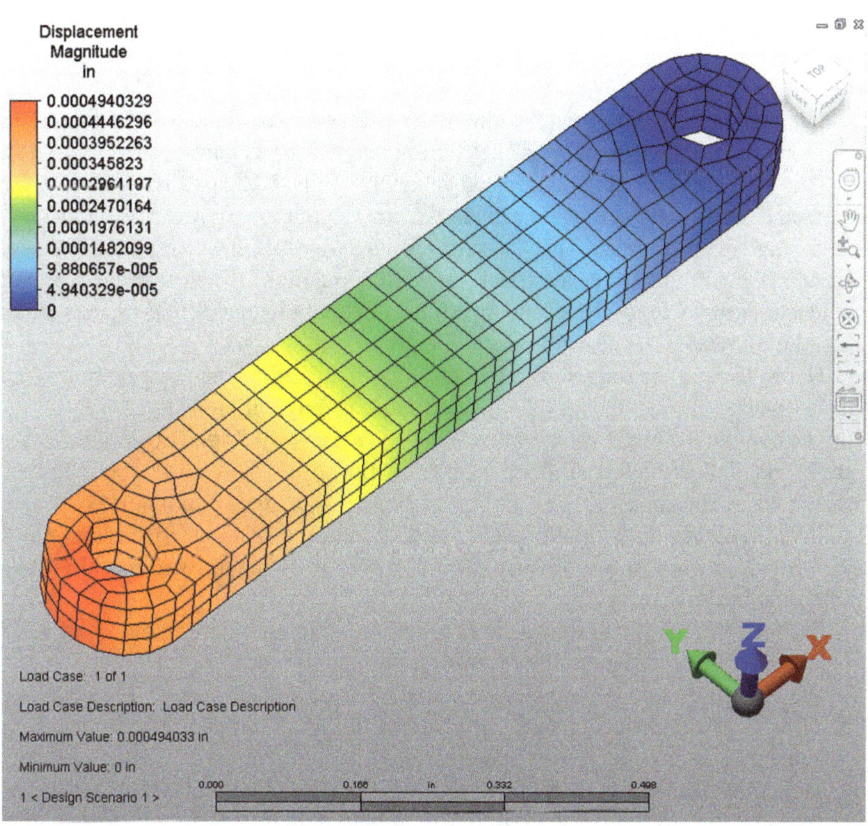

Fig. 10.31 Deflection profile of driving link

23,110.67 psi in this front end of internal shaft is less than the material yield strength of 36,000 psi and maximum deflection of 0.00144 in. is within material allowable deflection limit.

The following diagram shows the stress profile of internal driving shaft when surgeon delivers surgiclips distally during surgical procedures.

The computer-aided simulation and analysis in Figs. 10.16 and 10.17 present the stress and deflection of instrument internal shaft in this new biomedical open surgical instrument. The analytic results demonstrate that the maximum stress of 22,542.71 psi in this instrument internal shaft is less than the material yield strength of 36,000 psi and maximum deflection of 0.00152 in. is within material allowable deflection limit.

The following diagram shows the stress profile of instrument front jaw when surgeon forms the surgiclips during surgical procedures.

The computer-aided simulation and analysis in Figs. 10.19 and 10.20 indicate the stress and deflection of front jaw (closing tissue) in this new biomedical open surgical instrument. The analytic results state that the maximum stress of

16,300.16 psi in this front jaw (closing tissue) is less than the material yield strength of 36,000 psi and maximum deflection of 0.00421 in. is within material allowable deflection limit.

The following diagram shows the stress profile of instrument front jaw when surgeon manipulates the thick tissues during surgical procedures.

The computer-aided simulation and analysis in Figs. 10.21 and 10.22 display the stress and deflection of front jaw (thick tissue manipulation) in this new biomedical open surgical instrument. The analytic results show that the maximum stress of 23,995.81 psi in this front jaw (thick tissue manipulation) is less than the material yield strength of 36,000 psi and maximum deflection of 0.00875 in. is within material allowable deflection limit.

The following diagram shows the stress profile of instrument surgiclip driving pusher when surgeon delivers the surgiclip during surgical procedures.

The computer-aided simulation and analysis in Figs. 10.24 and 10.25 present the stress and deflection of surgiclip pusher in this new biomedical open surgical instrument. The analytic results demonstrate that the maximum stress of 21,389.27 psi in this surgiclip pusher is less than the material yield strength of 36,000 psi and maximum deflection of 0.03106 in. is within material allowable deflection limit.

The following diagram shows the stress profile of instrument external shaft when surgeon manipulates patient tissues during surgical procedures.

The computer-aided simulation and analysis in Figs. 10.27 and 10.28 indicate the stress and deflection of external shaft in this new biomedical open surgical instrument. The analytic results state that the maximum stress of 23,481.61 psi in this external shaft is less than the material yield strength of 36,000 psi and maximum deflection of 0.00675 in. is within material allowable deflection limit.

The following diagram shows the stress profile of instrument driver link when surgeon operates the device during surgical procedures.

The computer-aided simulation and analysis in Figs. 10.30 and 10.31 display the stress and deflection of driving link in this new biomedical open surgical instrument. The analytic results show that the maximum stress of 24,154.42 psi in this driving link is less than the material yield strength of 36,000 psi and maximum deflection of 0.00049 in. is within material allowable deflection limit.

The above computer-aided simulation results displayed that the maximum stresses on these critical components are all below the material yield stress and maximum material deflections are all within material allowable deformation limits. The above analytic solutions have confirmed the good function and reliable quality of this newly developed biomedical open surgiclip instrument.

10.2.2 Biomedical Endoscopic Surgiclip Instrument

The computer-aided kinematic simulations of multiple instrument setup have been performed on this new biomedical endoscopic surgiclip instrument to determine the optimal system design, with analytic result shown in Fig. 10.32.

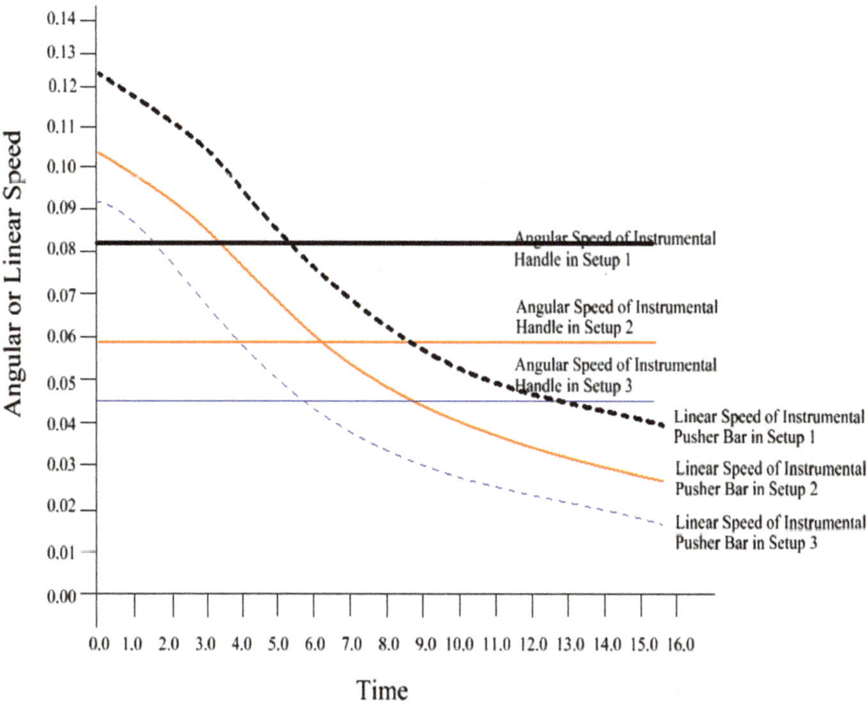

Fig. 10.32 Angular/linear speed vs. time in new biomedical endoscopic instrument

The multiple instrument setups have been analyzed based on the above computational simulation and the optimal system design can be achieved when mechanical advantage of this new instrument equals to 2.878, as shown in the following equation (White 2003):

$$MA = (VR) \times 2.218 = (0.04851/0.03738) \times 2.218 = 2.878 \tag{10.7}$$

This result indicates that when 20 lbf forces are required to fully close the biomedical endoscopic surgiclip, the surgeons only need to apply 3.475 lbf forces that are lower than the normal spec of 4 lbf in surgical operation procedure. This will benefit physicians and surgeons in the medical treatments and surgical operation procedures. Figures 10.33, 10.34, 10.35, 10.36, 10.37, 10.38, 10.39, 10.40, 10.41, 10.42, 10.43, 10.44, 10.45, 10.46, 10.47, 10.48, 10.49, 10.50, 10.51, 10.52, 10.53, 10.54, 10.55, 10.56, 10.57, 10.58, 10.59, 10.60, 10.61, 10.62, 10.63, 10.64, 10.65, 10.66, 10.67, 10.68, 10.69, 10.70, 10.71, 10.72, 10.73, 10.74, 10.75, and 10.76 display the stress and deflection profiles of major components in this newly developed biomedical endoscopic surgiclip instrument.

The computer-aided simulation and analysis in Figs. 10.34 and 10.35 tell the stress and deflection of surgiclip pusher in this new biomedical endoscopic surgical instrument. The analytic results present that the maximum stress of 12,113.39 psi in

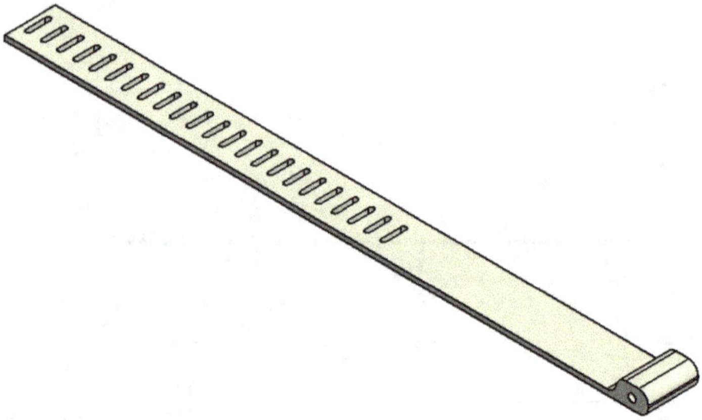

Fig. 10.33 Surgiclip pusher

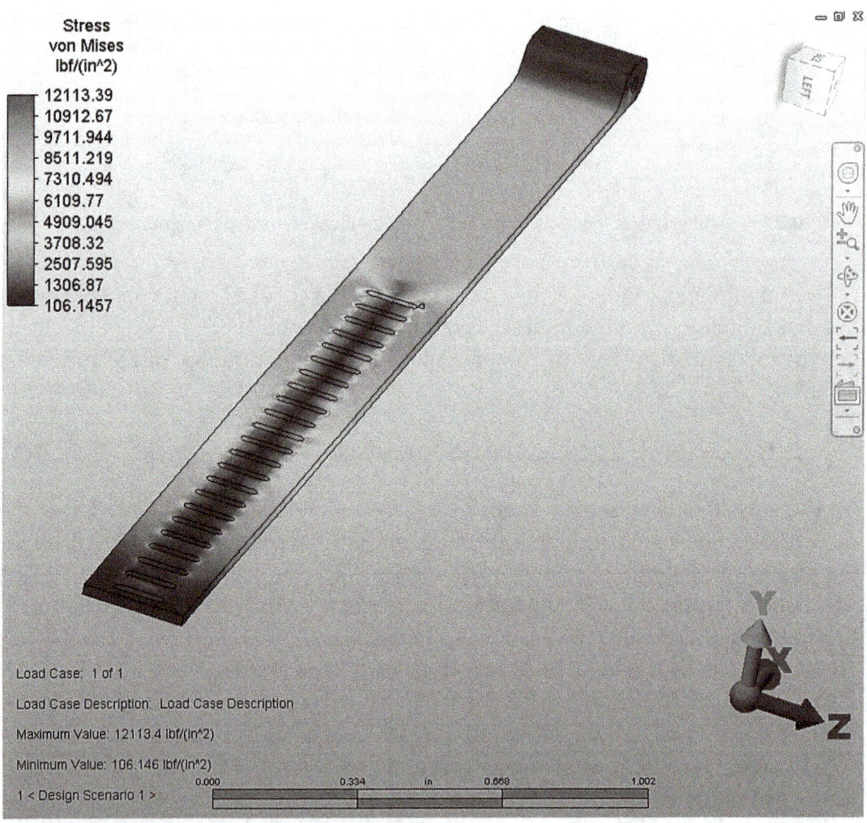

Fig. 10.34 Stress profile of surgiclip pusher

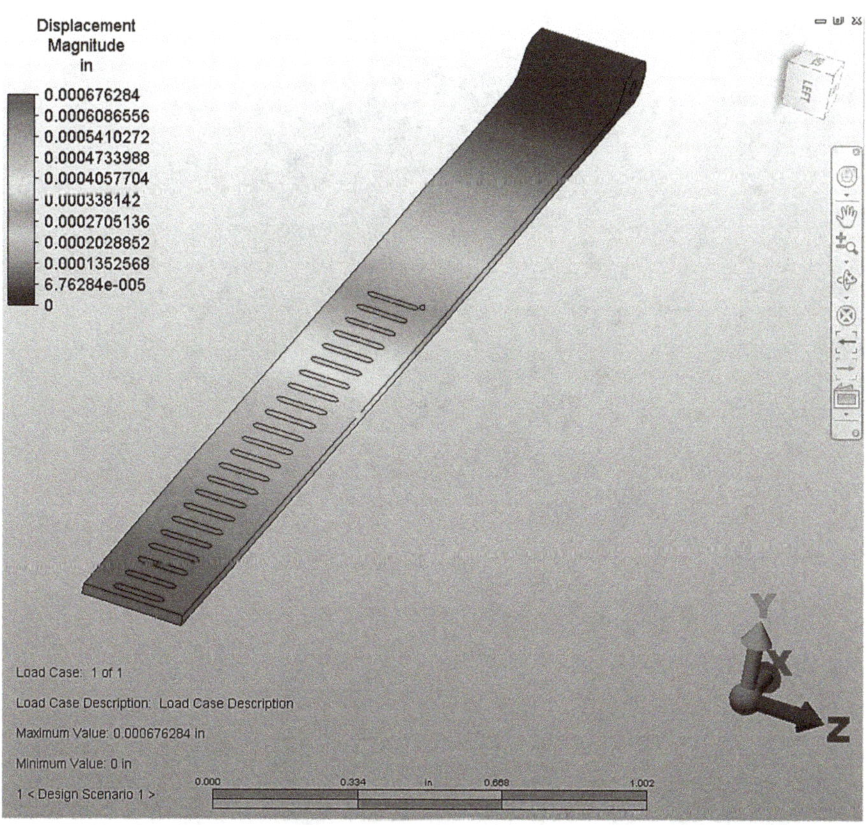

Fig. 10.35 Deflection profile of surgiclip pusher

Fig. 10.36 Gear_1

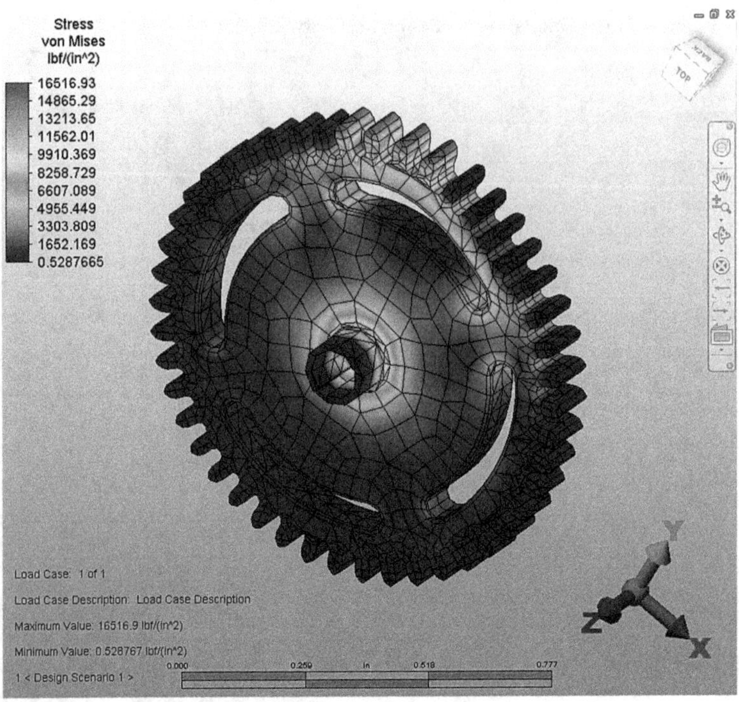

Fig. 10.37 Stress profile of gear_1

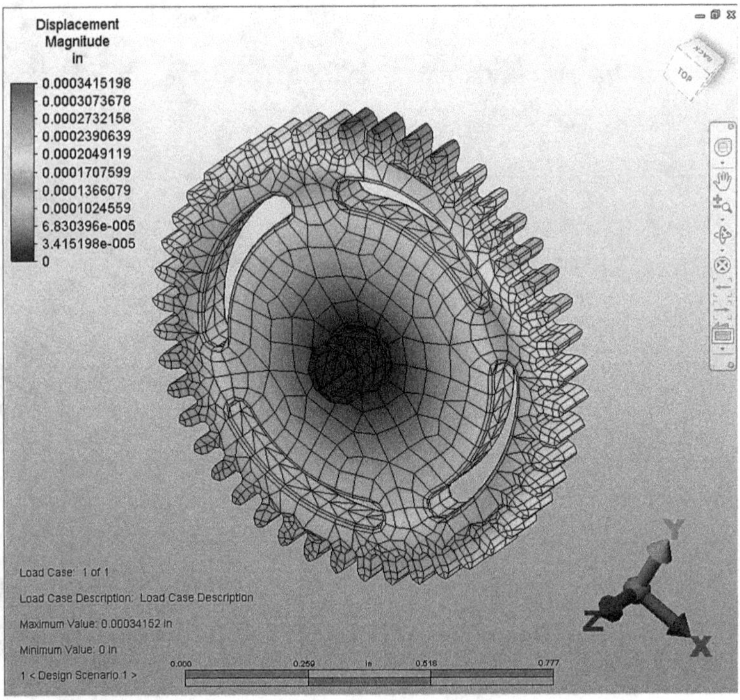

Fig. 10.38 Deflection profile of gear_1

Fig. 10.39 Gear_2

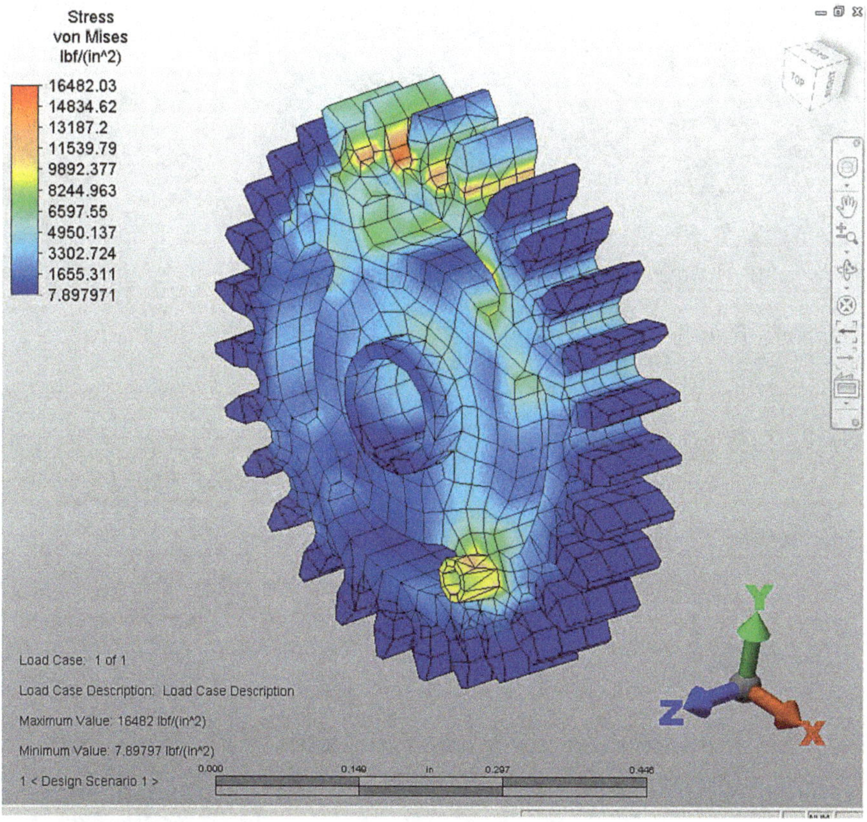

Fig. 10.40 Stress profile of gear_2

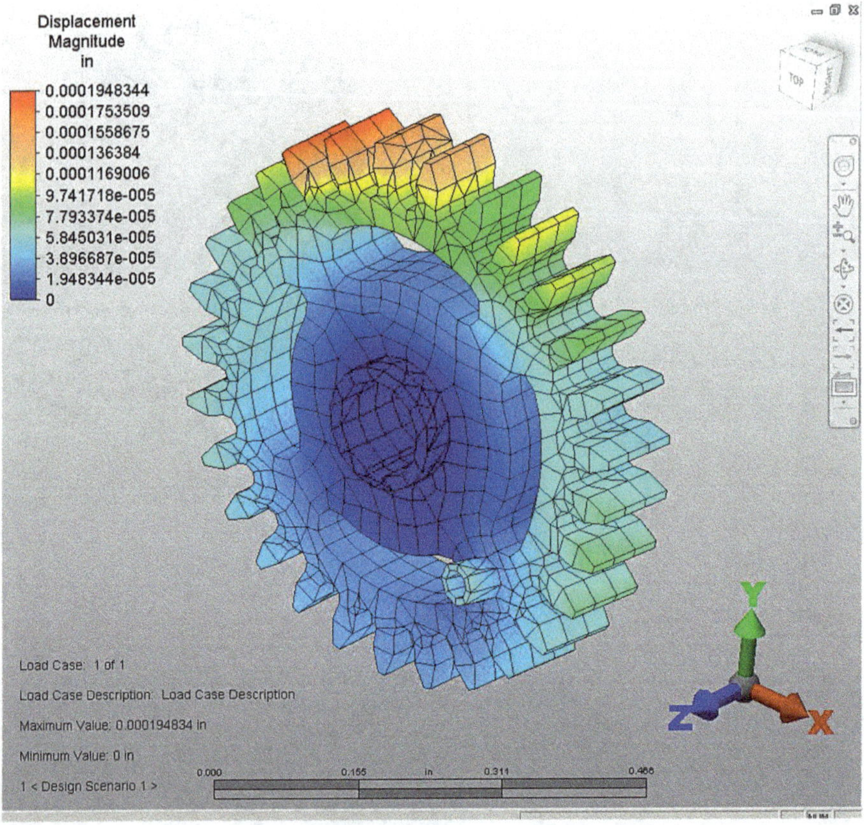

Fig. 10.41 Deflection profile of gear_2

Fig. 10.42 Gear_3

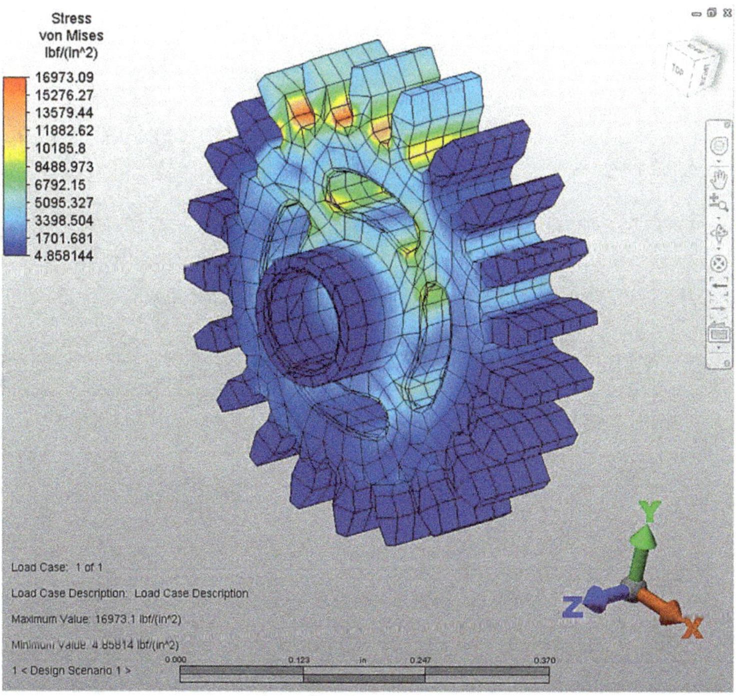

Fig. 10.43 Stress profile of gear_3

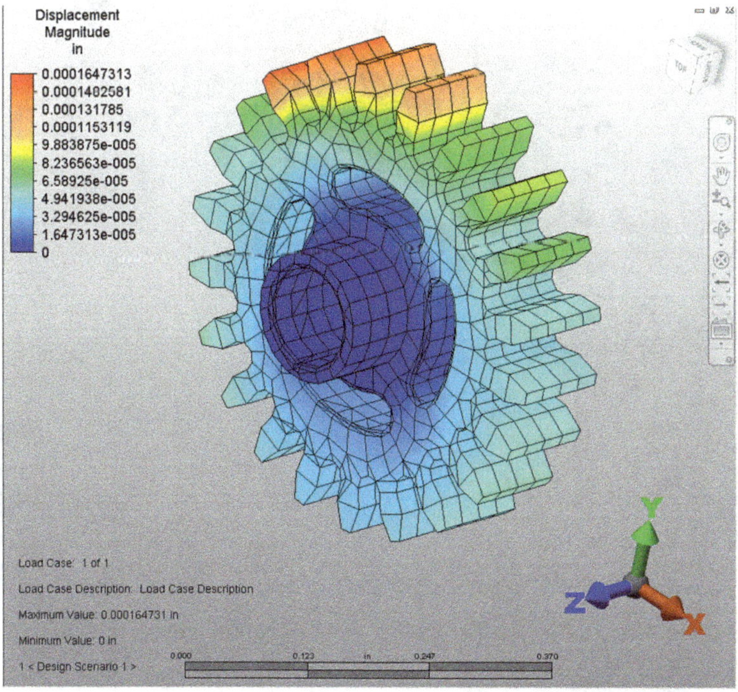

Fig. 10.44 Deflection profile of gear_3

Fig. 10.45 Gear_4

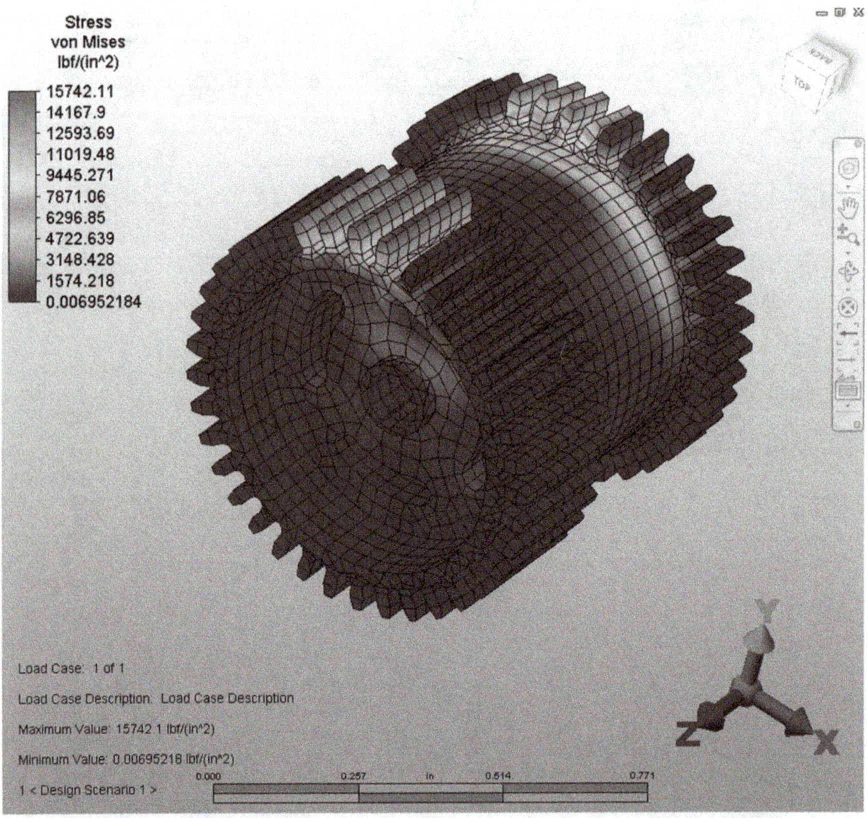

Fig. 10.46 Stress profile of gear_4

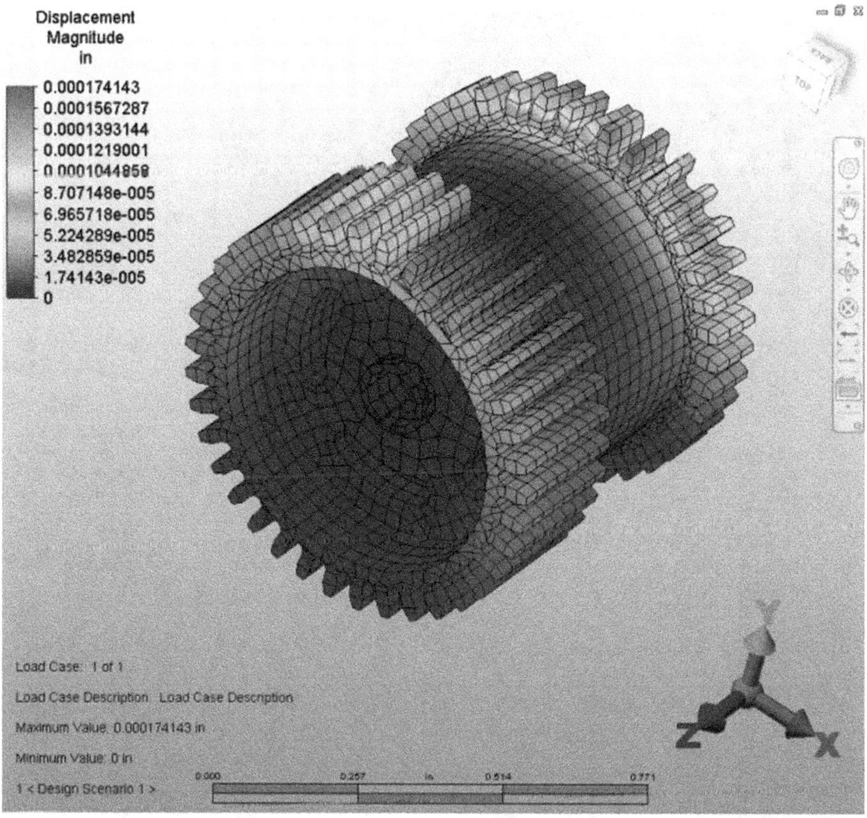

Fig. 10.47 Deflection profile of gear_4

Fig. 10.48 Rack gear

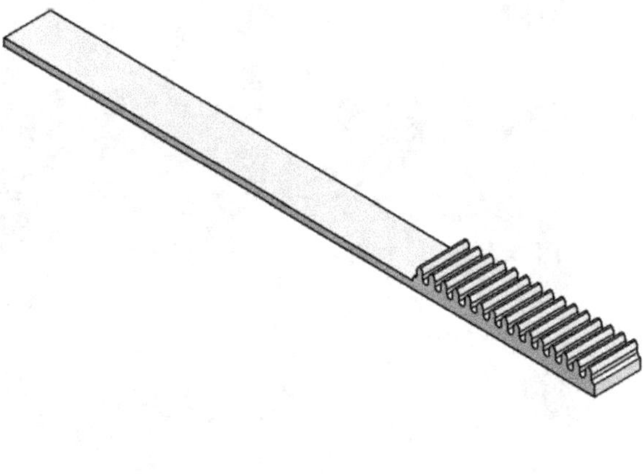

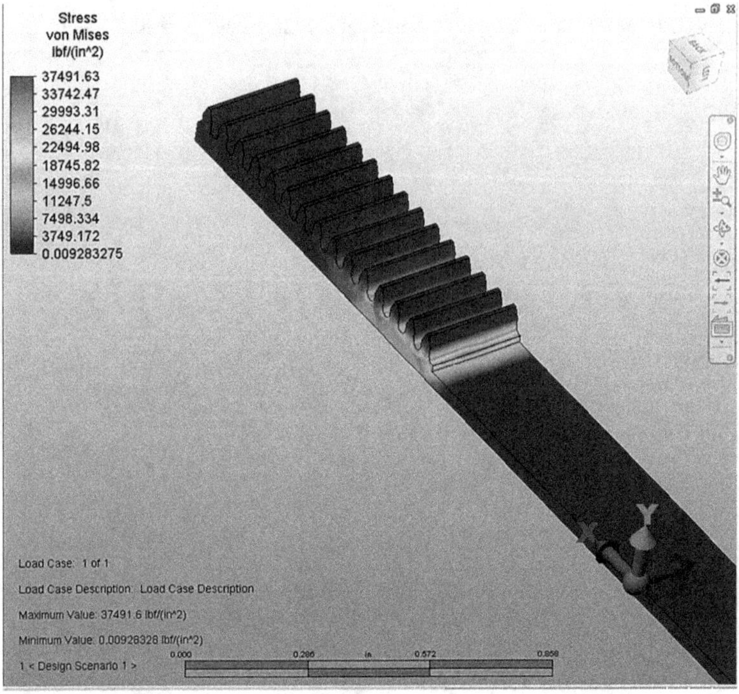

Fig. 10.49 Stress profile of rack gear

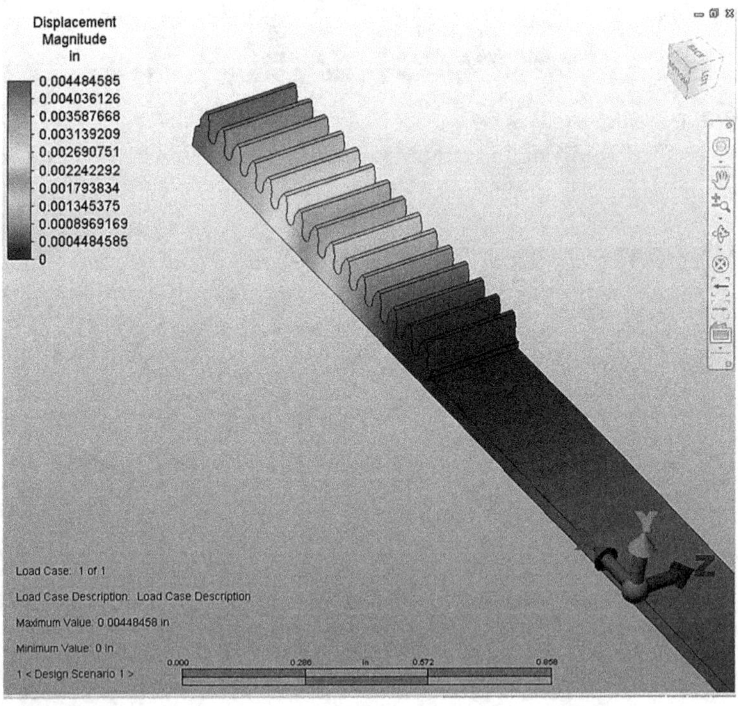

Fig. 10.50 Deflection profile of rack gear

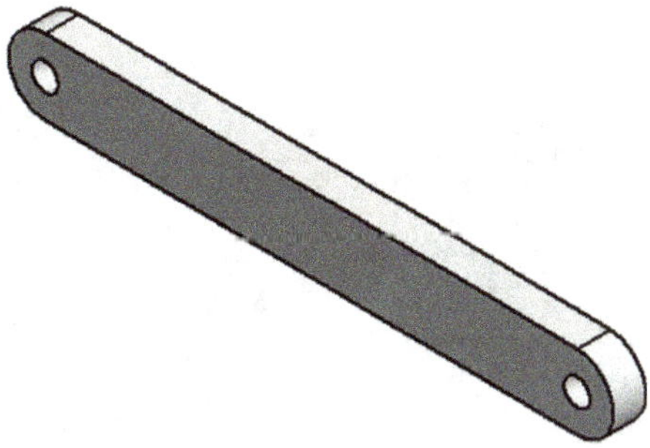

Fig. 10.51 Gear link

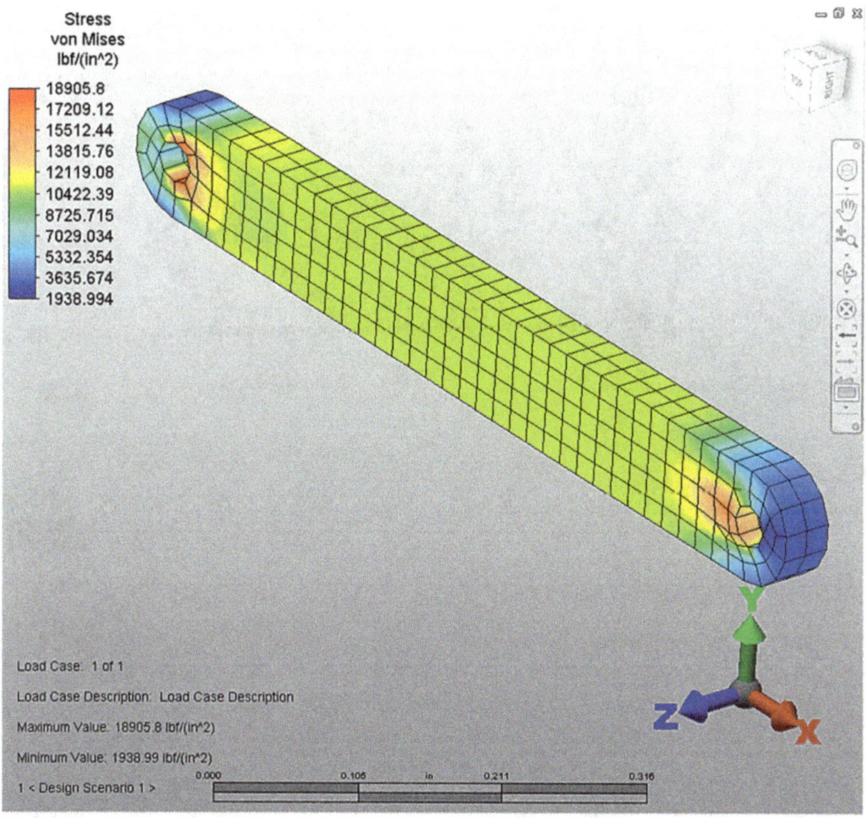

Fig. 10.52 Stress profile of gear link

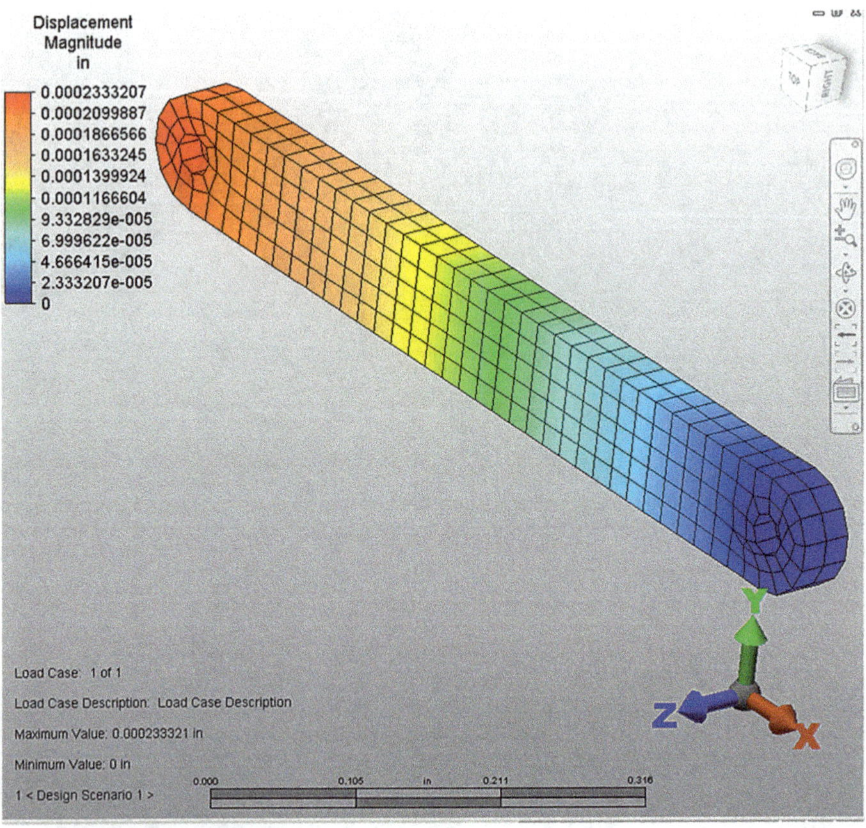

Fig. 10.53 Deflection profile of gear link

Fig. 10.54 Instrument handle trigger

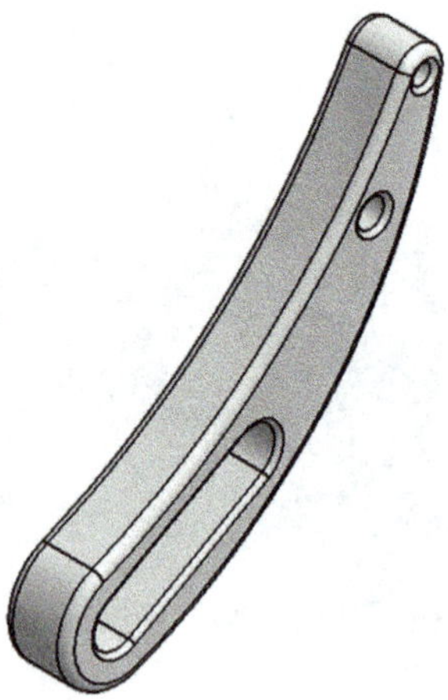

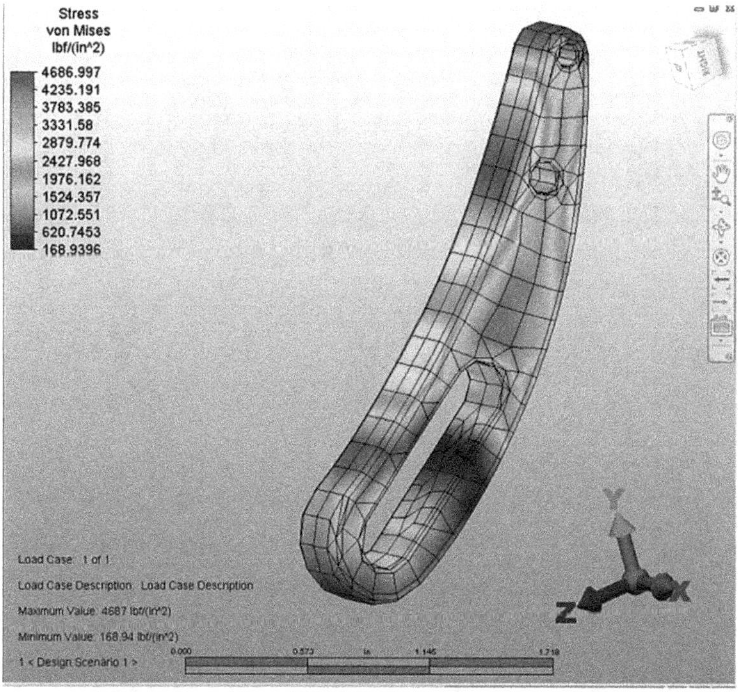

Fig. 10.55 Stress profile of instrument handle trigger

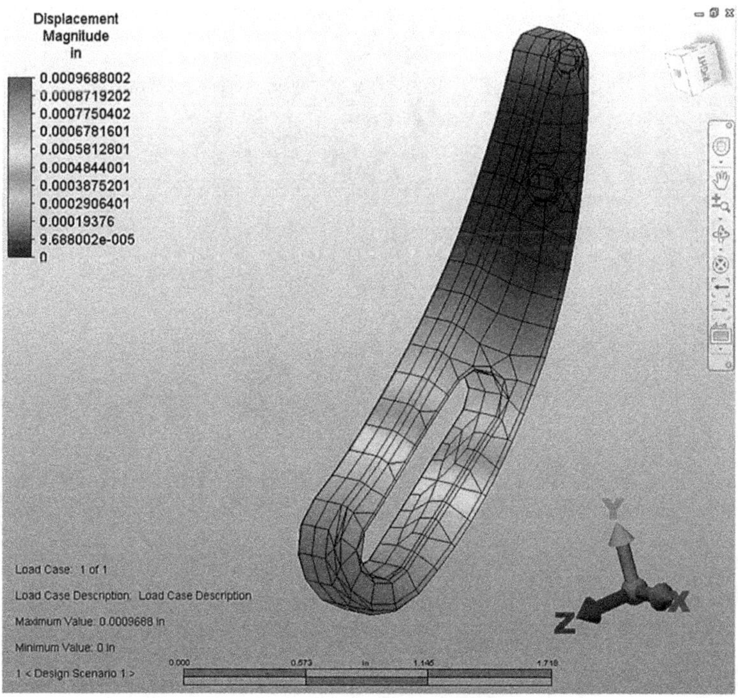

Fig. 10.56 Deflection profile of instrument handle trigger

Fig. 10.57 Instrument handle

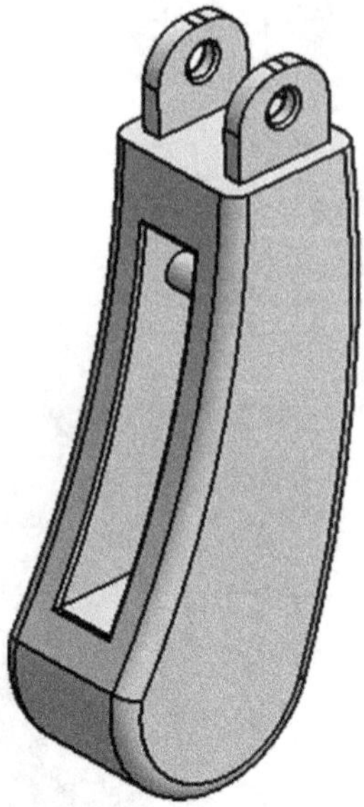

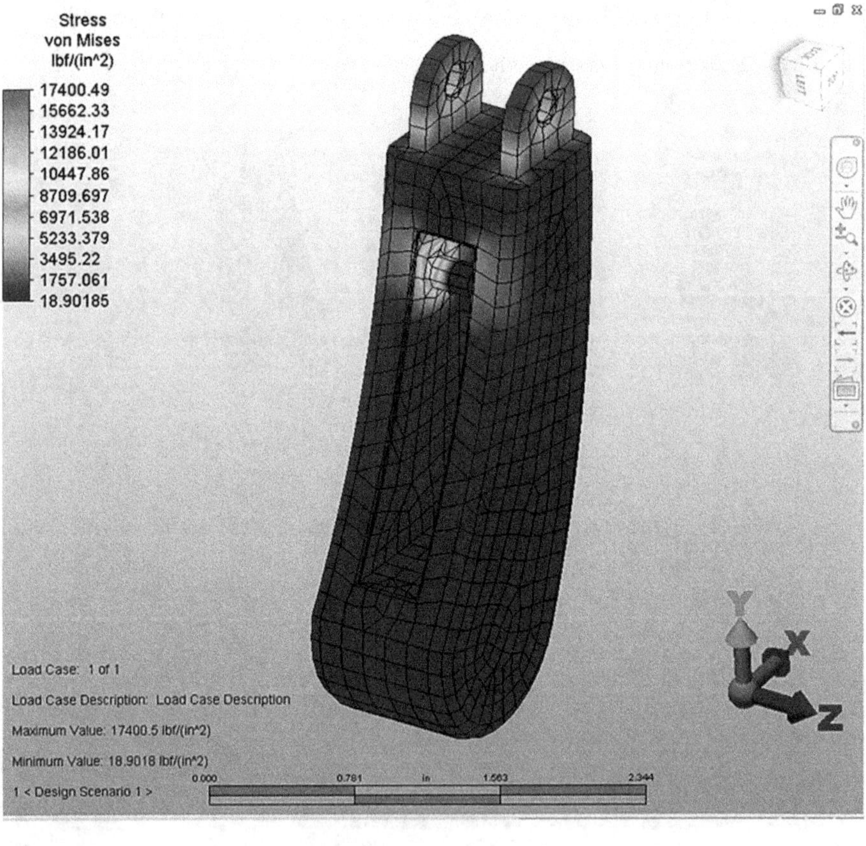

Fig. 10.58 Stress profile of instrument handle

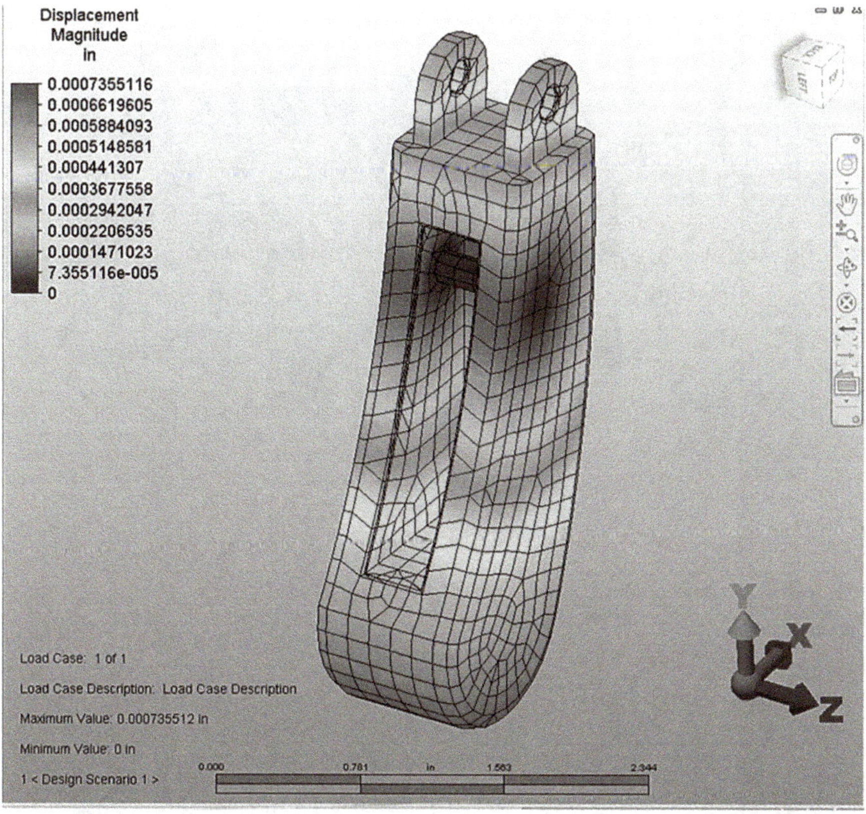

Fig. 10.59 Deflection profile of instrument handle

Fig. 10.60 Front end of inside shaft

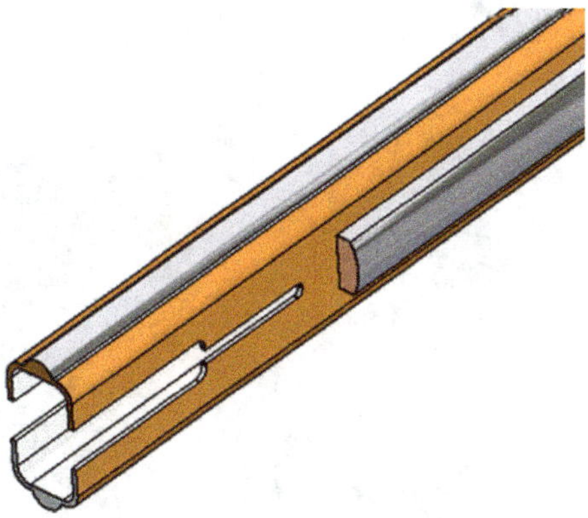

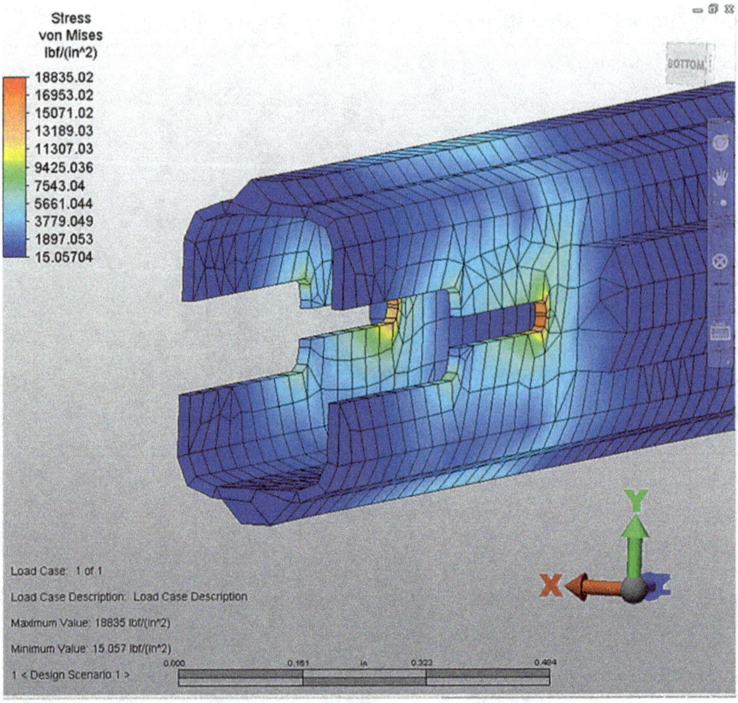

Fig. 10.61 Stress profile of front end in inside shaft

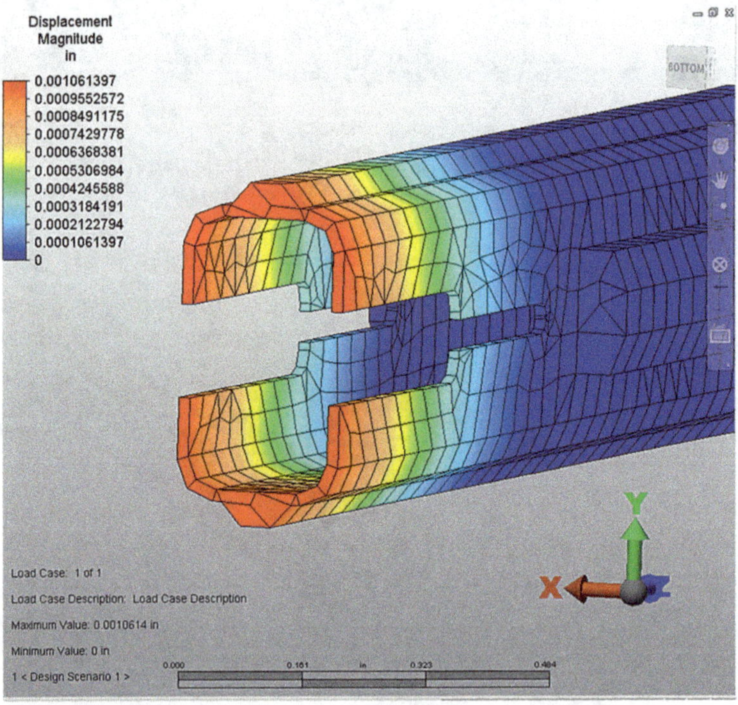

Fig. 10.62 Deflection profile of front end in inside shaft

Fig. 10.63 Inside shaft

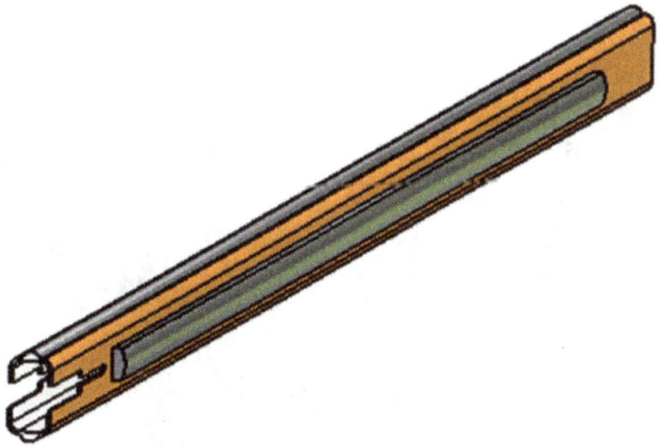

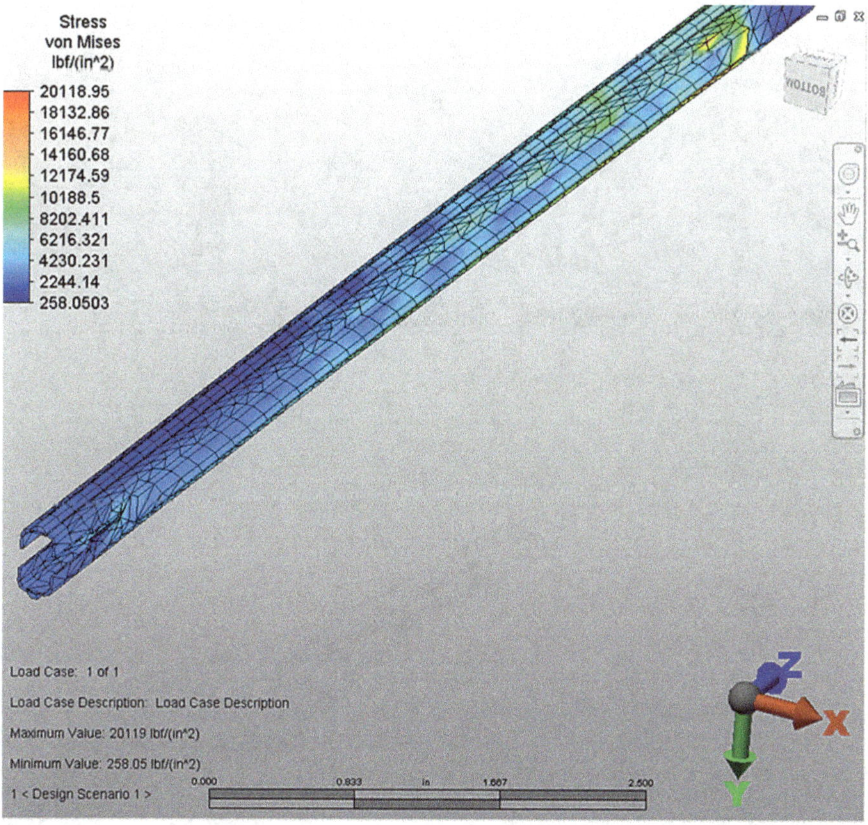

Fig. 10.64 Stress profile of inside shaft

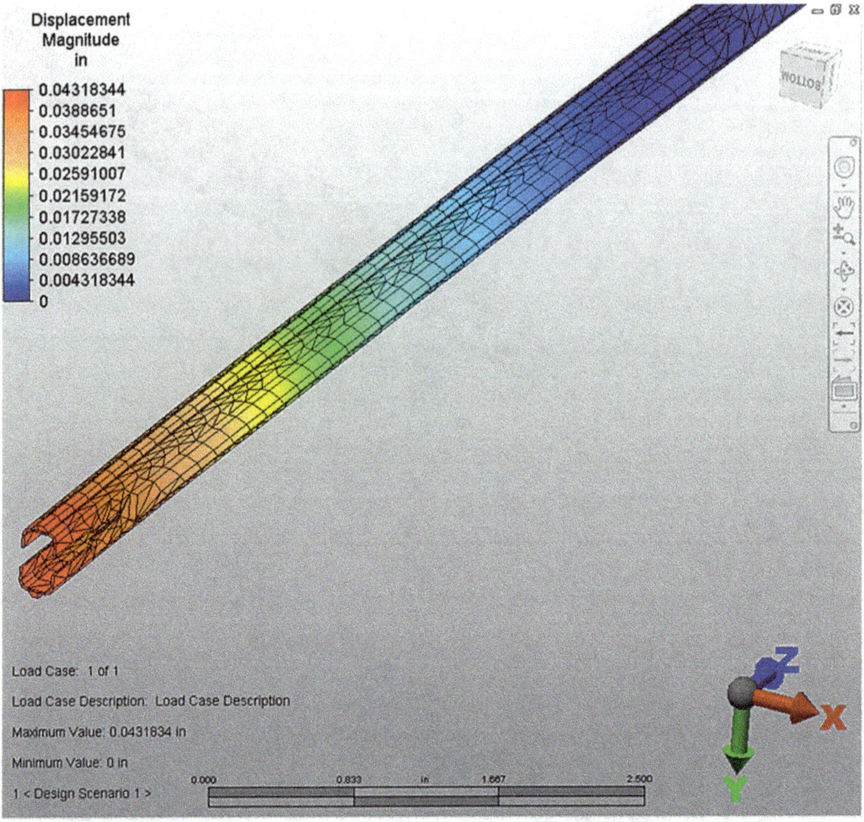

Fig. 10.65 Deflection profile of inside shaft

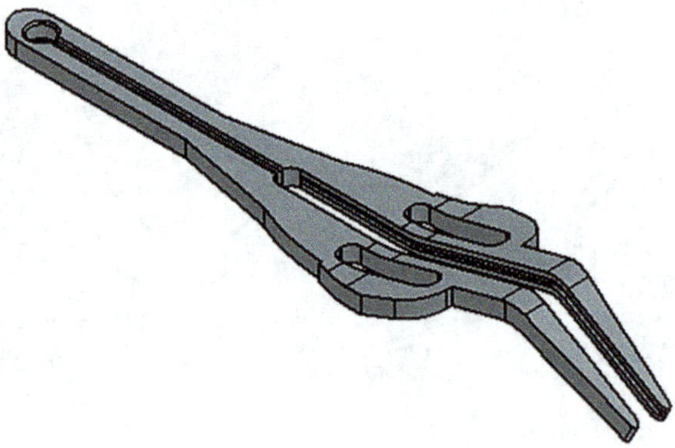

Fig. 10.66 Instrument front jaw

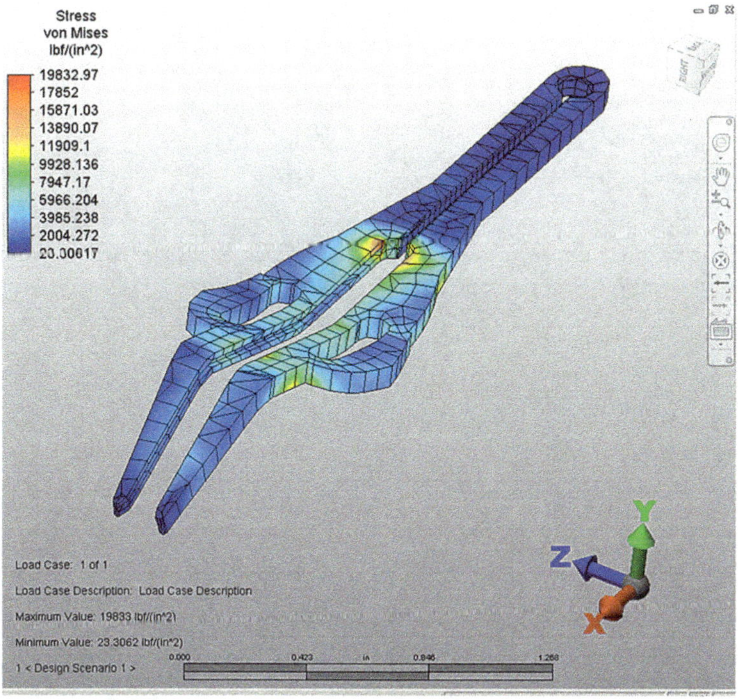

Fig. 10.67 Stress profile of jaw in closing the tissues

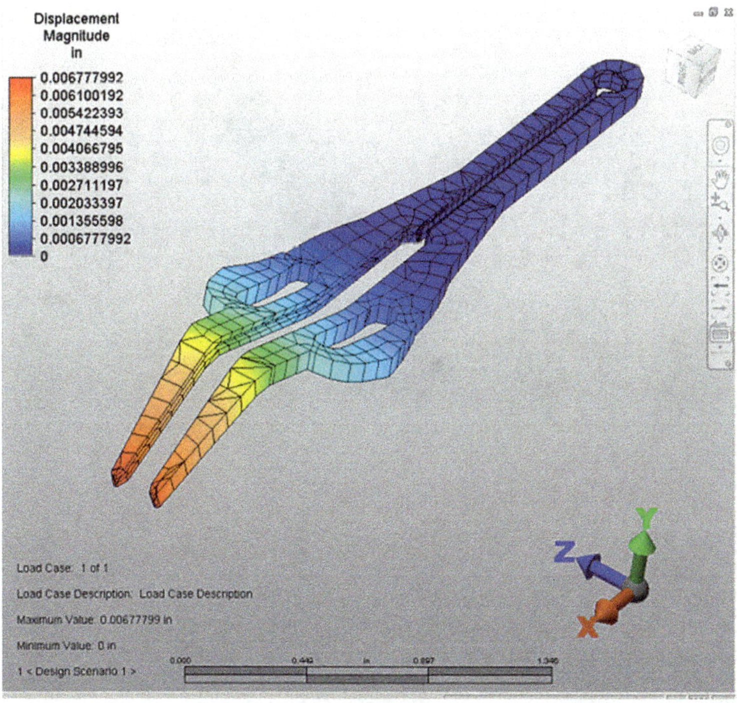

Fig. 10.68 Deflection profile of jaw in closing the tissues

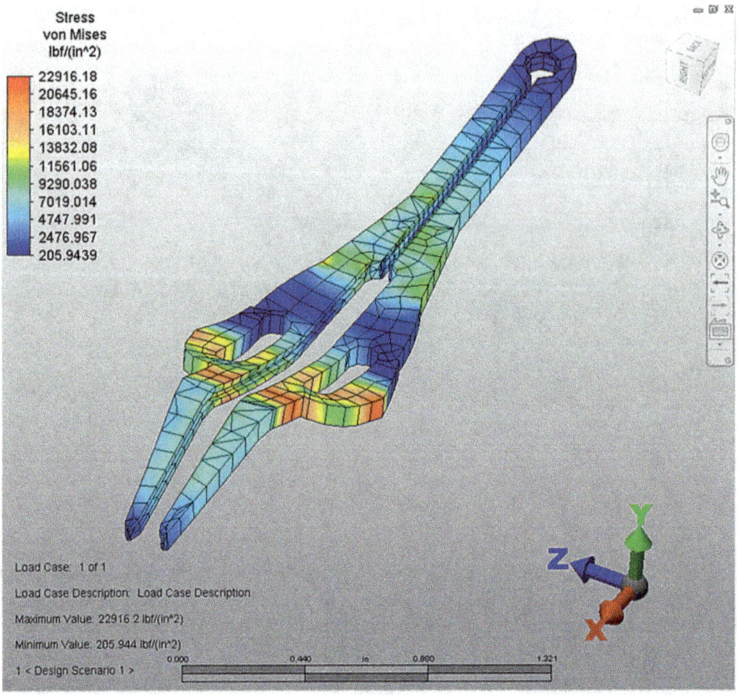

Fig. 10.69 Stress profile of instrument jaw in thick tissue manipulation

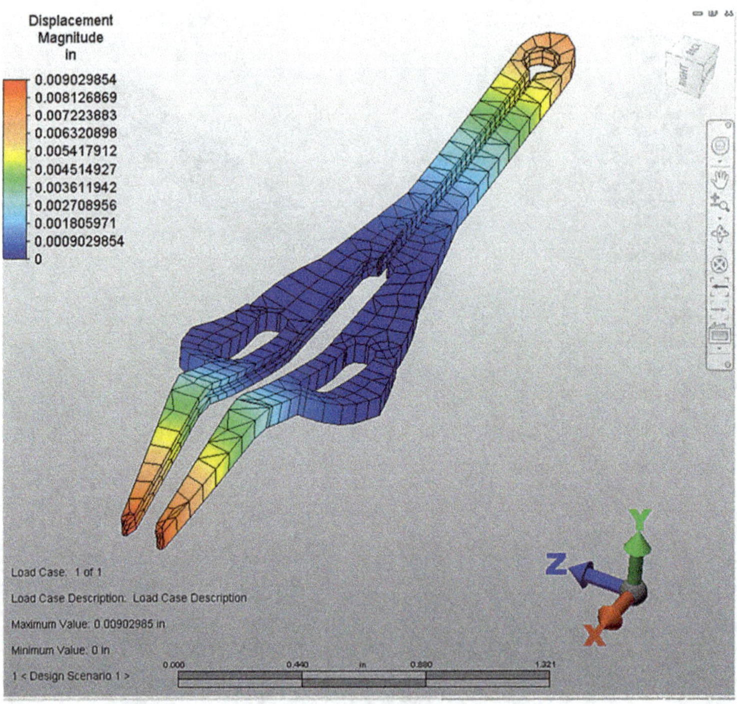

Fig. 10.70 Deflection profile of instrument jaw in thick tissue manipulation

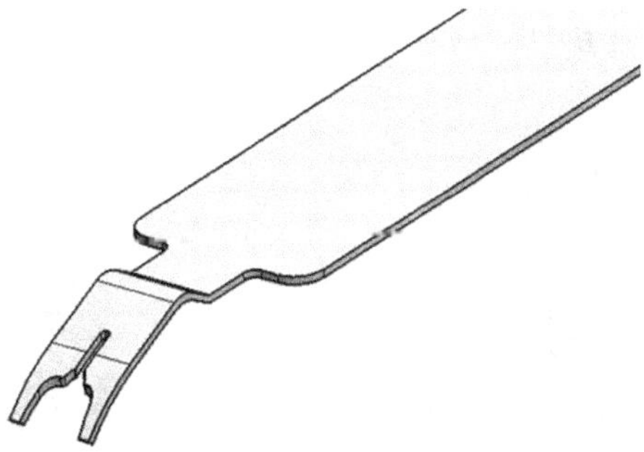

Fig. 10.71 Instrument surgiclip pusher

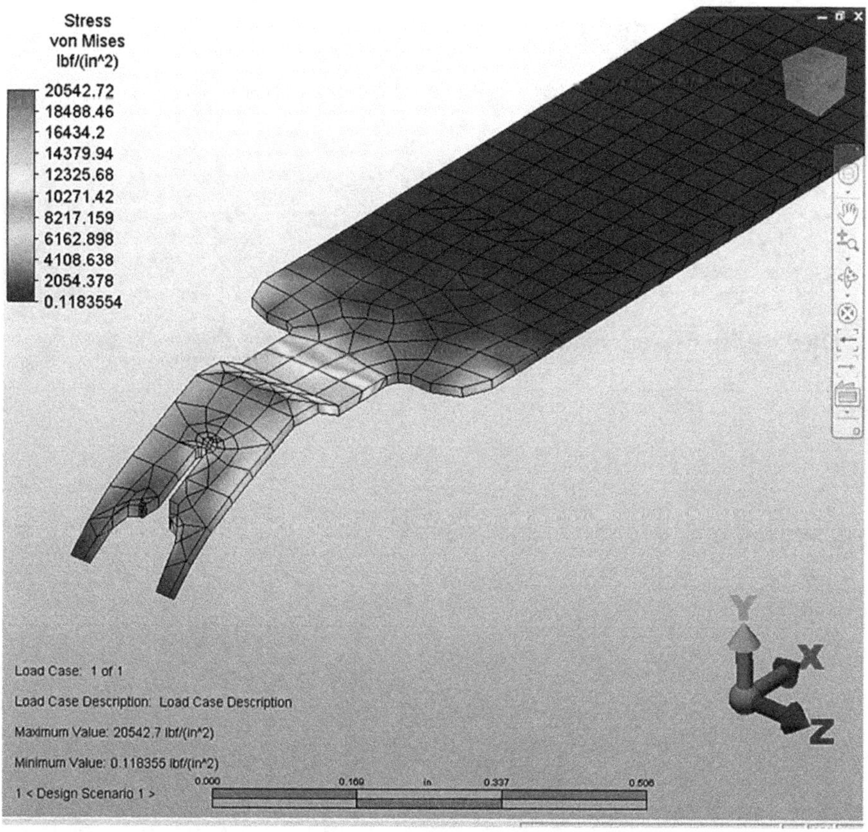

Fig. 10.72 Stress profile of instrument surgiclip pusher

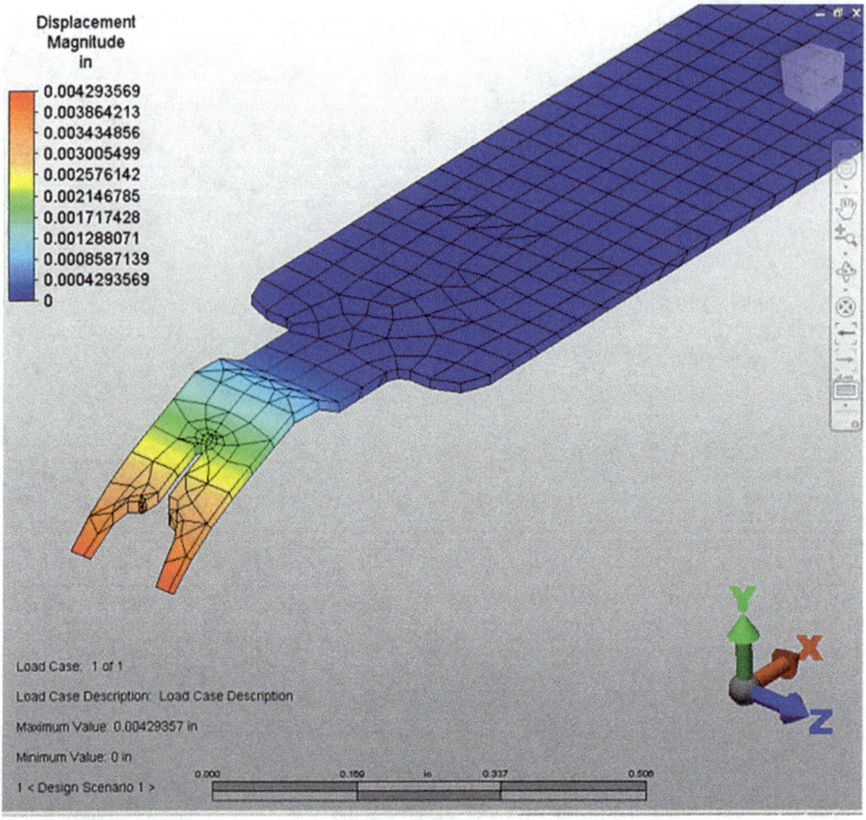

Fig. 10.73 Deflection profile of instrument surgiclip pusher

Fig. 10.74 External shaft

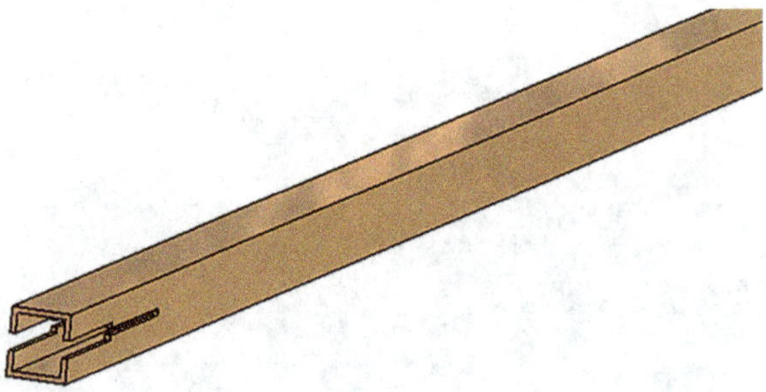

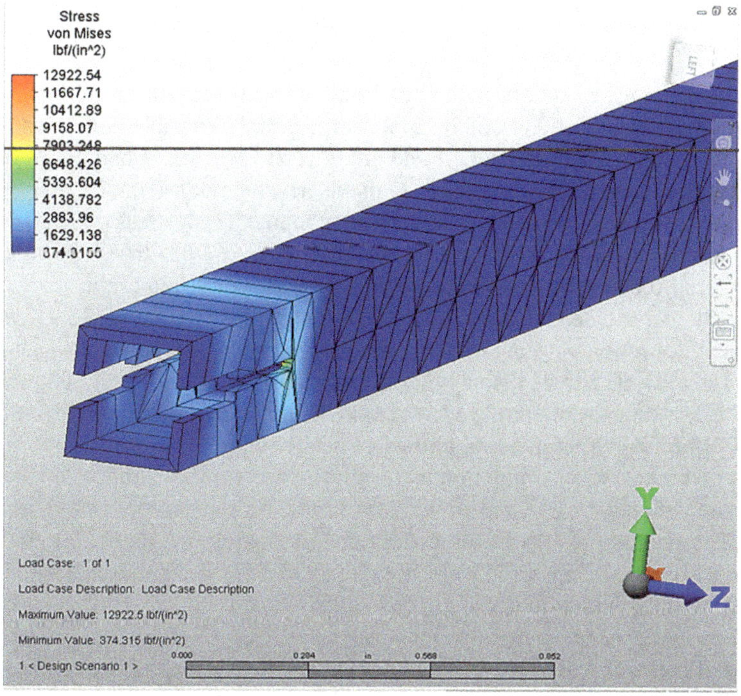

Fig. 10.75 Stress profile of external shaft

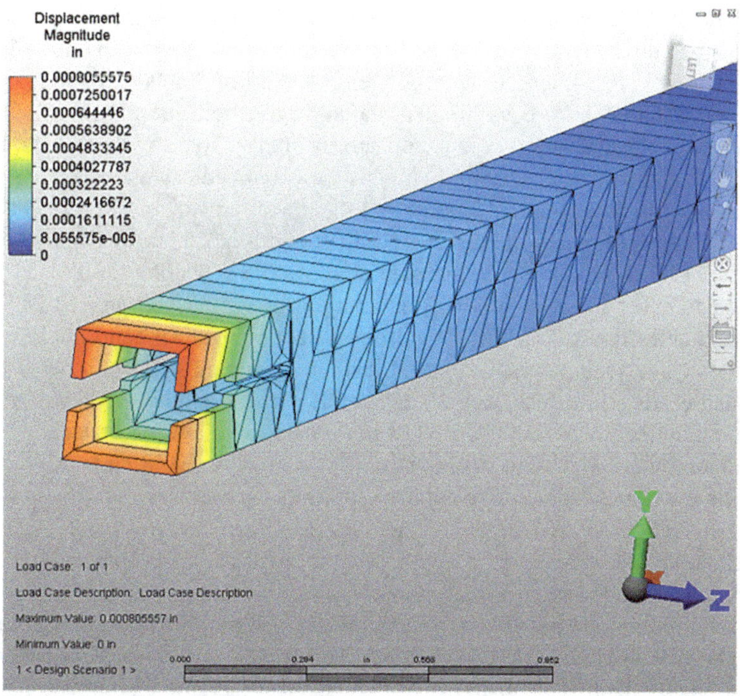

Fig. 10.76 Deflection profile of external shaft

this surgiclip pusher is less than the material yield strength of 36,000 psi and maximum deflection of 0.00068 in. is within material allowable deflection limit.

The computer-aided simulation and analysis in Figs. 10.37 and 10.38 demonstrate the stress and deflection of gear_1 in this new biomedical endoscopic surgical instrument. The analytic results indicate that the maximum stress of 16,516.93 psi in this gear_1 is less than the material yield strength of 36,000 psi and maximum deflection of 0.00034 in. is within material allowable deflection limit.

The computer-aided simulation and analysis in Figs. 10.40 and 10.41 state the stress and deflection of gear_2 in this new biomedical endoscopic surgical instrument. The analytic results display that the maximum stress of 16,482.03 psi in this gear_2 is less than the material yield strength of 36,000 psi and maximum deflection of 0.00019 in. is within material allowable deflection limit.

The computer-aided simulation and analysis in Figs. 10.43 and 10.44 show the stress and deflection of gear_3 in this new biomedical endoscopic surgical instrument. The analytic results tell that the maximum stress of 16,973.09 psi in this gear_3 is less than the material yield strength of 36,000 psi and maximum deflection of 0.00016 in. is within material allowable deflection limit.

The computer-aided simulation and analysis in Figs. 10.46 and 10.47 present the stress and deflection of gear_4 in this new biomedical endoscopic surgical instrument. The analytic results demonstrate that the maximum stress of 15,742.11 psi in this gear_4 is less than the material yield strength of 36,000 psi and maximum deflection of 0.00017 in. is within material allowable deflection limit.

The computer-aided simulation and analysis in Figs. 10.49 and 10.50 indicate the stress and deflection of rack gear in this new biomedical endoscopic surgical instrument. The analytic results state that the maximum stress of 37,491.63 psi in this rack gear is less than the material yield strength of 50,000 psi and maximum deflection of 0.00448 in. is within material allowable deflection limit.

The computer-aided simulation and analysis in Figs. 10.52 and 10.53 display the stress and deflection of gear link in this new biomedical endoscopic surgical instrument. The analytic results show that the maximum stress of 18,905.80 psi in this gear link is less than the material yield strength of 36,000 psi and maximum deflection of 0.00023 in. is within material allowable deflection limit.

The computer-aided simulation and analysis in Figs. 10.55 and 10.58 tell the stress and deflection of handle trigger in this new biomedical endoscopic surgical instrument. The analytic results present that the maximum stress of 4,686.99 psi in this handle trigger is less than the material yield strength of 36,000 psi and maximum deflection of 0.00097 in. is within material allowable deflection limit.

The computer-aided simulation and analysis in Figs. 10.58 and 10.59 demonstrate the stress and deflection of handle in this new biomedical endoscopic surgical instrument. The analytic results indicate that the maximum stress of 17,400.49 psi in this handle is less than the material yield strength of 36,000 psi and maximum deflection of 0.00074 in. is within material allowable deflection limit.

The computer-aided simulation and analysis in Figs. 10.61 and 10.62 state the stress and deflection of front end of inside shaft in this new biomedical endoscopic surgical instrument. The analytic results display that the maximum stress of

18,835.02 psi in this front end of inside shaft is less than the material yield strength of 36,000 psi and maximum deflection of 0.00106 in. is within material allowable deflection limit.

The computer-aided simulation and analysis in Figs. 10.64 and 10.65 show the stress and deflection of inside shaft in this new biomedical endoscopic surgical instrument. The analytic results tell that the maximum stress of 18,835.02 psi in this inside shaft is less than the material yield strength of 36,000 psi and maximum deflection of 0.00106 in. is within material allowable deflection limit.

The computer-aided simulation and analysis in Figs. 10.67 and 10.68 present the stress and deflection of jaw (closing tissue) in this new biomedical endoscopic surgical instrument. The analytic results demonstrate that the maximum stress of 19,832.97 psi in this jaw (closing tissue) is less than the material yield strength of 36,000 psi and maximum deflection of 0.00678 in. is within material allowable deflection limit.

The computer-aided simulation and analysis in Figs. 10.69 and 10.70 indicate the stress and deflection of jaw (thick tissue manipulation) in this new biomedical endoscopic surgical instrument. The analytic results state that the maximum stress of 22,916.18 psi in this jaw (thick tissue manipulation) is less than the material yield strength of 36,000 psi and maximum deflection of 0.00903 in. is within material allowable deflection limit.

The computer-aided simulation and analysis in Figs. 10.72 and 10.73 display the stress and deflection of surgiclip pusher in this new biomedical endoscopic surgical instrument. The analytic results show that the maximum stress of 20,542.72 psi in this surgiclip pusher is less than the material yield strength of 36,000 psi and maximum deflection of 0.00429 in. is within material allowable deflection limit.

The computer-aided simulation and analysis in Figs. 10.75 and 10.76 tell the stress and deflection of external shaft in this new biomedical endoscopic surgical instrument. The analytic results present that the maximum stress of 12,922.54 psi in this external shaft is less than the material yield strength of 36,000 psi and maximum deflection of 0.00081 in. is within material allowable deflection limit.

These computational simulation results show that the maximum stresses on these critical components are all less than the material yield stress and maximum material deflections are all within material allowable deformation limits. The computational solutions confirm that this newly developed biomedical endoscopic surgiclip instrument can function properly and reliably to assist doctors and surgeons in their medical treatments and surgical operations.

10.3 Experiment on Biomedical and Surgical Systems

Two newly designed biomedical and surgical systems are being prototyped and tested to compare and verify the results from computer-aided simulation.

Table 10.1 Prototype testing of driving gear in this new biomedical open surgiclip instrument

Number of experiment	Maximum stress (psi)	Maximum deflection (in.)
1	23,663.45	0.00071
2	23,663.38	0.00068
3	23,663.76	0.00065
4	23,663.75	0.00078
5	23,663.69	0.00069
6	23,663.78	0.00065
7	23,663.24	0.00067
8	23,663.29	0.00068
9	23,663.59	0.00063
10	23,663.79	0.00061
11	23,663.88	0.00062
12	23,663.97	0.00065
13	23,663.98	0.00061
14	23,663.54	0.00062
15	23,663.84	0.00063
16	23,663.95	0.00062
17	23,663.66	0.00061
18	23,663.88	0.00063
19	23,663.59	0.00062
20	23,663.65	0.00061
Average	23,663.68	0.00065

10.3.1 Experiment on Biomedical Open Surgiclip Instrument

Table 10.1 records the prototype testing results of driving gear in this new biomedical open surgiclip instrument.

The prototype experimental results of driving gear in Table 10.1 verify the proper function because the average maximum stress 23,663.68 psi and average maximum deflection 0.00065 in. are approximately equal to the results of maximum stress 23,663.32 psi and maximum deflection 0.00077 in. that are laid out, respectively, in Figs. 10.7 and 10.8 by computer-aided modeling and numerical simulation.

Table 10.2 lays out the prototype testing results of trigger in this new biomedical open surgiclip instrument.

The prototype experimental results of trigger in Table 10.2 confirm the appropriate function since the average maximum stress 1,380.64 psi and average maximum deflection 0.01920 in. are approximately same as the results of maximum stress 1,380.27 psi and maximum deflection 0.01944 in. that are displayed, respectively, in Figs. 10.10 and 10.11 by computer-aided modeling and numerical simulation.

Table 10.3 shows the prototype testing results of instrument front end of internal shaft in this new biomedical open surgiclip instrument.

Table 10.2 Prototype testing of trigger in this new biomedical open surgiclip instrument

Number of experiment	Maximum stress (psi)	Maximum deflection (in.)
1	1,380.37	0.01924
2	1,380.48	0.01918
3	1,380.77	0.01948
4	1,380.78	0.01915
5	1,380.25	0.01924
6	1,380.24	0.01918
7	1,380.59	0.01916
8	1,380.78	0.01922
9	1,380.98	0.01921
10	1,380.89	0.01924
11	1,380.22	0.01908
12	1,380.21	0.01911
13	1,380.78	0.01912
14	1,380.84	0.01916
15	1,380.85	0.01924
16	1,380.77	0.01918
17	1,380.88	0.01917
18	1,380.78	0.01919
19	1,380.65	0.01921
20	1,380.67	0.01922
Average	1,380.64	0.01920

Table 10.3 Prototype testing of instrument front end of internal shaft in this new biomedical open surgiclip instrument

Number of experiment	Maximum stress (psi)	Maximum deflection (in.)
1	23,110.47	0.00154
2	23,110.48	0.00148
3	23,110.38	0.00159
4	23,110.39	0.00154
5	23,110.68	0.00157
6	23,110.55	0.00159
7	23,110.54	0.00156
8	23,110.39	0.00155
9	23,110.35	0.00159
10	23,110.48	0.00157
11	23,110.43	0.00159
12	23,110.35	0.00155
13	23,110.38	0.00157
14	23,110.36	0.00156
15	23,110.33	0.00159
16	23,110.31	0.00157
17	23,110.41	0.00159
18	23,110.32	0.00156
19	23,110.44	0.00155
20	23,110.48	0.00154
Average	23,110.43	0.00156

Table 10.4 Prototype testing of instrument internal shaft in this new biomedical open surgiclip instrument

Number of experiment	Maximum stress (psi)	Maximum deflection (in.)
1	22,542.61	0.00159
2	22,542.52	0.00155
3	22,542.48	0.00148
4	22,542.57	0.00149
5	22,542.38	0.00159
6	22,542.39	0.00157
7	22,542.62	0.00161
8	22,542.55	0.00162
9	22,542.54	0.00164
10	22,542.61	0.00159
11	22,542.54	0.00165
12	22,542.48	0.00163
13	22,542.54	0.00162
14	22,542.62	0.00165
15	22,542.36	0.00164
16	22,542.38	0.00165
17	22,542.39	0.00164
18	22,542.44	0.00165
19	22,542.49	0.00163
20	22,542.52	0.00161
Average	22,542.50	0.00161

The prototype experimental results for instrument front end of internal shaft in Table 10.3 prove the normal function as the average maximum stress 23,110.43 psi and average maximum deflection 0.00156 in. are similar to the results of maximum stress 23,110.67 psi and maximum deflection 0.00144 in. that are exhibited, respectively, in Figs. 10.13 and 10.14 by computer-aided modeling and numerical simulation.

Table 10.4 presents the prototype testing results of instrument internal shaft in this new biomedical open surgiclip instrument.

The prototype experimental results of instrument internal shaft in Table 10.4 verify the proper function because the average maximum stress 22,542.50 psi and average maximum deflection 0.00161 in. are almost equal to the results of maximum stress 22,542.71 psi and maximum deflection 0.00152 in. that are depicted, respectively, in Figs. 10.16 and 10.17 by computer-aided modeling and numerical simulation.

Table 10.5 represents the prototype testing results of instrument jaw when fully closing surgiclip to secure the tissues by using this new biomedical open surgiclip instrument.

The prototype experimental results of instrument jaw (to finally form the surgiclip to secure the tissues) in Table 10.5 confirm the appropriate function since the average maximum stress 18,300.47 psi and average maximum deflection

Table 10.5 Prototype testing of instrument jaw when fully closing surgiclip to secure the tissues by using this new biomedical open surgiclip instrument

Number of experiment	Maximum stress (psi)	Maximum deflection (in.)
1	18,300.25	0.00424
2	18,300.24	0.00429
3	18,300.11	0.00433
4	18,300.54	0.00438
5	18,300.48	0.00439
6	18,300.39	0.00435
7	18,300.38	0.00444
8	18,300.54	0.00448
9	18,300.55	0.00449
10	18,300.45	0.00446
11	18,300.56	0.00445
12	18,300.49	0.00448
13	18,300.48	0.00449
14	18,300.55	0.00418
15	18,300.65	0.00421
16	18,300.57	0.00447
17	18,300.56	0.00449
18	18,300.59	0.00446
19	18,300.48	0.00436
20	18,300.45	0.00438
Average	18,300.47	0.00439

0.00439 in. are very similar to the results of maximum stress 18,300.15 psi and maximum deflection 0.00421 in. that are depicted, respectively, in Figs. 10.19 and 10.20 by computer-aided modeling and numerical simulation.

Table 10.6 expresses the prototype testing results of instrument jaw in thick tissue manipulation by using this new biomedical open surgiclip instrument.

The prototype experimental results of instrument jaw (to manipulate the thick tissue) in Table 10.6 prove the normal function as the average maximum stress 23,995.52 psi and average maximum deflection 0.00863 in. are closely equal to the results of maximum stress 23,995.81 psi and maximum deflection 0.00875 in. that are indicated, respectively, in Figs. 10.21 and 10.22 by computer-aided modeling and numerical simulation.

Table 10.7 records the prototype testing results of instrument surgiclip pusher in this new biomedical open surgiclip instrument.

The prototype experimental results of instrument surgiclip pusher in Table 10.7 verify the proper function because the average maximum stress 21,389.46 psi and average maximum deflection 0.03121 in. are almost same as the results of maximum stress 21,389.27 psi and maximum deflection 0.03106 in. that are laid out, respectively, in Figs. 10.24 and 10.25 by computer-aided modeling and numerical simulation.

Table 10.6 Prototype testing of instrument jaw in thick tissue manipulation by using this new biomedical open surgiclip instrument

Number of experiment	Maximum stress (psi)	Maximum deflection (in.)
1	23,995.75	0.00866
2	23,995.71	0.00862
3	23,995.55	0.00861
4	23,995.59	0.00876
5	23,995.48	0.00871
6	23,995.44	0.00868
7	23,995.42	0.00862
8	23,995.38	0.00861
9	23,995.39	0.00862
10	23,995.49	0.00863
11	23,995.54	0.00865
12	23,995.57	0.00861
13	23,995.44	0.00860
14	23,995.48	0.00861
15	23,995.54	0.00860
16	23,995.51	0.00863
17	23,995.55	0.00862
18	23,995.48	0.00861
19	23,995.51	0.00860
20	23,995.54	0.00862
Average	23,995.52	0.00863

Table 10.7 Prototype testing of instrument surgiclip pusher in this new biomedical open surgiclip instrument

Number of experiment	Maximum stress (psi)	Maximum deflection (in.)
1	21,389.37	0.03108
2	21,389.38	0.03112
3	21,389.29	0.03118
4	21,389.39	0.03116
5	21,389.48	0.03115
6	21,389.54	0.03124
7	21,389.52	0.03125
8	21,389.51	0.03118
9	21,389.50	0.03125
10	21,389.49	0.03124
11	21,389.48	0.03126
12	21,389.44	0.03122
13	21,389.47	0.03123
14	21,389.46	0.03125
15	21,389.39	0.03121
16	21,389.54	0.03124
17	21,389.50	0.03125
18	21,389.51	0.03124
19	21,389.48	0.03120
20	21,389.44	0.03118
Average	21,389.46	0.03121

Table 10.8 Prototype testing of instrument external shaft in this new biomedical open surgiclip instrument

Number of experiment	Maximum stress (psi)	Maximum deflection (in.)
1	23,481.55	0.00684
2	23,481.51	0.00688
3	23,481.48	0.00695
4	23,481.55	0.00699
5	23,481.44	0.00698
6	23,481.42	0.00695
7	23,481.38	0.00699
8	23,481.24	0.00697
9	23,481.35	0.00698
10	23,481.37	0.00688
11	23,481.41	0.00690
12	23,481.24	0.00689
13	23,481.29	0.00684
14	23,481.35	0.00686
15	23,481.38	0.00689
16	23,481.36	0.00695
17	23,481.33	0.00698
18	23,481.38	0.00695
19	23,481.44	0.00689
20	23,481.42	0.00691
Average	23,481.39	0.00692

Table 10.8 states the prototype testing results of instrument external shaft in this new biomedical open surgiclip instrument.

The prototype experimental results of instrument external shaft in Table 10.8 confirm the appropriate function since the average maximum stress 23,481.39 psi and average maximum deflection 0.00692 in. are approximately equal to the results of maximum stress 23,481.61 psi and maximum deflection 0.00675 in. that are shown, respectively, in Figs. 10.27 and 10.28 by computer-aided modeling and numerical simulation.

Table 10.9 demonstrates the prototype testing results of instrument driving link in this new biomedical open surgiclip instrument.

The prototype experimental results of instrument driving link in Table 10.9 prove the normal function as the average maximum stress 24,154.17 psi and average maximum deflection 0.00056 in. are approximately same as the results of maximum stress 24,154.42 psi and maximum deflection 0.00049 in. that are presented, respectively, in Figs. 10.30 and 10.31 by computer-aided modeling and numerical simulation.

Table 10.9 Prototype testing of instrument driving link in this new biomedical open surgiclip instrument

Number of experiment	Maximum stress (psi)	Maximum deflection (in.)
1	24,154.33	0.00050
2	24,154.22	0.00048
3	24,154.12	0.00055
4	24,154.11	0.00057
5	24,154.18	0.00059
6	24,154.08	0.00055
7	24,154.22	0.00057
8	24,154.08	0.00059
9	24,154.07	0.00055
10	24,154.12	0.00057
11	24,154.09	0.00056
12	24,154.10	0.00059
13	24,154.32	0.00054
14	24,154.17	0.00055
15	24,154.24	0.00057
16	24,154.15	0.00059
17	24,154.18	0.00057
18	24,154.20	0.00056
19	24,154.17	0.00059
20	24,154.19	0.00057
Average	24,154.17	0.00056

10.3.2 Experiment on Biomedical Endoscopic Surgiclip System

Table 10.10 demonstrates the prototype testing results of instrument surgiclip pusher in this new biomedical endoscopic surgiclip instrument.

The prototype experimental results of instrument surgiclip pusher in Table 10.10 verify the proper function because the average maximum stress 12,113.70 psi and average maximum deflection 0.00084 in. are close to the results of maximum stress 12,113.39 psi and maximum deflection 0.00068 in. that are represented, respectively, in Figs. 10.34 and 10.35 by computer-aided modeling and numerical simulation.

Table 10.11 expresses the prototype testing results of instrument gear_1 in this new biomedical endoscopic surgiclip instrument.

The prototype experimental results of instrument gear_1 in Table 10.11 confirm the appropriate function since the average maximum stress 16,516.39 psi and average maximum deflection 0.00051 in. are similar to the results of maximum stress 16,516.93 psi and maximum deflection 0.00034 in. that are indicated, respectively, in Figs. 10.36 and 10.37 by computer-aided modeling and numerical simulation.

Table 10.10 Prototype testing of instrument surgiclip pusher in this new biomedical endoscopic surgiclip instrument

Number of experiment	Maximum stress (psi)	Maximum deflection (in.)
1	12,113.38	0.00066
2	12,114.35	0.00069
3	12,113.85	0.00078
4	12,113.78	0.00082
5	12,113.39	0.00097
6	12,114.15	0.00065
7	12,114.18	0.00088
8	12,113.16	0.00093
9	12,113.18	0.00084
10	12,113.89	0.00082
11	12,113.98	0.00095
12	12,113.76	0.00098
13	12,113.24	0.00086
14	12,113.22	0.00084
15	12,113.78	0.00092
16	12,113.32	0.00095
17	12,114.38	0.00095
18	12,113.88	0.00088
19	12,113.99	0.00065
20	12,113.18	0.00069
Average	12,113.70	0.00084

Table 10.11 Prototype testing of instrument gear_1 in this new biomedical endoscopic surgiclip instrument

Number of experiment	Maximum stress (psi)	Maximum deflection (in.)
1	16,516.77	0.00039
2	16,516.95	0.00041
3	16,516.17	0.00045
4	16,516.08	0.00049
5	16,516.98	0.00054
6	16,516.38	0.00032
7	16,516.18	0.00057
8	16,515.98	0.00055
9	16,515.38	0.00059
10	16,516.12	0.00055
11	16,517.15	0.00059
12	16,516.02	0.00057
13	16,516.17	0.00059
14	16,516.38	0.00054
15	16,516.98	0.00059
16	16,516.24	0.00059
17	16,516.18	0.00057
18	16,516.01	0.00048
19	16,516.99	0.00039
20	16,516.78	0.00036
Average	16,516.39	0.00051

Table 10.12 Prototype testing of instrument gear_2 in this new biomedical endoscopic surgiclip instrument

Number of experiment	Maximum stress (psi)	Maximum deflection (in.)
1	16,482.13	0.00022
2	16,481.84	0.00024
3	16,481.38	0.00038
4	16,482.03	0.00032
5	16,481.84	0.00038
6	16,481.35	0.00036
7	16,481.11	0.00035
8	16,481.12	0.00038
9	16,482.33	0.00032
10	16,481.33	0.00039
11	16,481.45	0.00037
12	16,481.08	0.00026
13	16,482.01	0.00029
14	16,482.05	0.00039
15	16,481.24	0.00038
16	16,481.15	0.00039
17	16,481.38	0.00035
18	16,481.48	0.00029
19	16,482.15	0.00037
20	16,482.01	0.00025
Average	16,481.62	0.00033

Table 10.12 records the prototype testing results of instrument gear_2 in this new biomedical endoscopic surgiclip instrument.

The prototype experimental results of instrument gear_2 in Table 10.12 prove the normal function as the average maximum stress 16,481.62 psi and average maximum deflection 0.00033 in. are almost equal to the results of maximum stress 16,482.03 psi and maximum deflection 0.00019 in. that are laid out, respectively, in Figs. 10.40 and 10.41 by computer-aided modeling and numerical simulation.

Table 10.13 states the prototype testing results of instrument gear_3 in this new biomedical endoscopic surgiclip instrument.

The prototype experimental results of instrument gear_3 in Table 10.13 verify the proper function because the average maximum stress 16,973.55 psi and average maximum deflection 0.00030 in. are very similar to the results of maximum stress 16,973.09 psi and maximum deflection 0.00016 in. that are shown, respectively, in Figs. 10.43 and 10.44 by computer-aided modeling and numerical simulation.

Table 10.14 demonstrates the prototype testing results of instrument gear_4 in this new biomedical endoscopic surgiclip instrument.

The prototype experimental results of instrument gear_4 in Table 10.14 confirm the appropriate function since the average maximum stress 15,741.63 psi and average maximum deflection 0.00029 in. are closely equal to the results of maximum stress 15,742.11 psi and maximum deflection 0.00017 in. that are presented,

Table 10.13 Prototype testing of instrument gear_3 in this new biomedical endoscopic surgiclip instrument

Number of experiment	Maximum stress (psi)	Maximum deflection (in.)
1	16,973.24	0.00018
2	16,973.02	0.00015
3	16,973.89	0.00027
4	16,973.78	0.00019
5	16,973.65	0.00033
6	16,973.05	0.00038
7	16,973.88	0.00039
8	16,973.79	0.00024
9	16,973.99	0.00029
10	16,973.01	0.00039
11	16,973.97	0.00037
12	16,973.95	0.00038
13	16,973.04	0.00018
14	16,973.89	0.00037
15	16,973.88	0.00036
16	16,973.86	0.00039
17	16,973.03	0.00038
18	16,973.88	0.00037
19	16,973.19	0.00015
20	16,973.07	0.00019
Average	16,973.55	0.00030

Table 10.14 Prototype testing of instrument gear_4 in this new biomedical endoscopic surgiclip instrument

Number of experiment	Maximum stress (psi)	Maximum deflection (in.)
1	15,742.01	0.00019
2	15,742.08	0.00015
3	15,741.38	0.00024
4	15,741.15	0.00029
5	15,742.14	0.00018
6	15,741.09	0.00025
7	15,742.15	0.00033
8	15,741.24	0.00038
9	15,741.05	0.00039
10	15,742.18	0.00018
11	15,741.24	0.00027
12	15,741.25	0.00039
13	15,741.18	0.00035
14	15,741.48	0.00037
15	15,742.12	0.00027
16	15,742.11	0.00038
17	15,741.24	0.00039
18	15,741.38	0.00033
19	15,742.02	0.00029
20	15,742.19	0.00015
Average	15,741.63	0.00029

Table 10.15 Prototype testing of instrument gear rack in this new biomedical endoscopic surgiclip instrument

Number of experiment	Maximum stress (psi)	Maximum deflection (in.)
1	37,491.24	0.00438
2	37,491.08	0.00412
3	37,491.83	0.00418
4	37,490.33	0.00454
5	37,490.84	0.00402
6	37,491.18	0.00408
7	37,491.05	0.00411
8	37,491.21	0.00449
9	37,491.12	0.00403
10	37,491.88	0.00412
11	37,491.06	0.00405
12	37,491.11	0.00402
13	37,490.33	0.00408
14	37,491.02	0.00416
15	37,491.14	0.00417
16	37,491.24	0.00419
17	37,491.73	0.00455
18	37,490.55	0.00402
19	37,491.78	0.00438
20	37,491.54	0.00451
Average	37,491.16	0.00421

respectively, in Figs. 10.46 and 10.47 by computer-aided modeling and numerical simulation.

Table 10.15 demonstrates the prototype testing results of instrument gear rack in this new biomedical endoscopic surgiclip instrument.

The prototype experimental results of instrument gear rack in Table 10.15 prove the normal function as the average maximum stress 37,491.16 psi and average maximum deflection 0.00421 in. are almost same as the results of maximum stress 37,491.63 psi and maximum deflection 0.00448 in. that are represented, respectively, in Figs. 10.49 and 10.50 by computer-aided modeling and numerical simulation.

Table 10.16 expresses the prototype testing results of instrument gear link in this new biomedical endoscopic surgiclip instrument.

The prototype experimental results of instrument gear link in Table 10.16 verify the proper function because the average maximum stress 18,905.29 psi and average maximum deflection 0.00032 in. are approximately equal to the results of maximum stress 18,905.80 psi and maximum deflection 0.00023 in. that are indicated, respectively, in Figs. 10.52 and 10.53 by computer-aided modeling and numerical simulation.

Table 10.17 records the prototype testing results of instrument handle trigger in this new biomedical endoscopic surgiclip instrument.

Table 10.16 Prototype testing of instrument gear link in this new biomedical endoscopic surgiclip instrument

Number of experiment	Maximum stress (psi)	Maximum deflection (in.)
1	18,905.25	0.00024
2	18,905.78	0.00029
3	18,904.55	0.00038
4	18,904.88	0.00037
5	18,904.85	0.00039
6	18,905.90	0.00027
7	18,905.48	0.00025
8	18,904.68	0.00021
9	18,904.88	0.00038
10	18,905.18	0.00037
11	18,905.38	0.00035
12	18,905.08	0.00029
13	18,905.24	0.00039
14	18,905.84	0.00029
15	18,905.12	0.00022
16	18,905.16	0.00038
17	18,905.98	0.00037
18	18,905.18	0.00039
19	18,905.44	0.00029
20	18,905.98	0.00021
Average	18,905.29	0.00032

Table 10.17 Prototype testing of instrument handle trigger in this new biomedical endoscopic surgiclip instrument

Number of experiment	Maximum stress (psi)	Maximum deflection (in.)
1	4,686.98	0.00092
2	4,686.78	0.00099
3	4,686.09	0.00091
4	4,686.12	0.00088
5	4,686.24	0.00084
6	4,686.11	0.00081
7	4,687.19	0.00092
8	4,686.38	0.00083
9	4,686.05	0.00082
10	4,686.18	0.00084
11	4,686.38	0.00081
12	4,686.54	0.00083
13	4,687.29	0.00092
14	4,687.24	0.00081
15	4,686.48	0.00085
16	4,686.12	0.00082
17	4,686.08	0.00081
18	4,686.11	0.00091
19	4,686.77	0.00082
20	4,687.26	0.00093
Average	4,686.52	0.00086

Table 10.18 Prototype testing of instrument handle in this new biomedical endoscopic surgiclip instrument

Number of experiment	Maximum stress (psi)	Maximum deflection (in.)
1	17,400.48	0.00072
2	17,400.79	0.00071
3	17,400.42	0.00068
4	17,401.09	0.00066
5	17,401.55	0.00065
6	17,400.88	0.00071
7	17,401.68	0.00078
8	17,400.99	0.00065
9	17,400.77	0.00063
10	17,400.98	0.00061
11	17,400.89	0.00062
12	17,400.84	0.00064
13	17,400.24	0.00062
14	17,400.48	0.00064
15	17,400.99	0.00068
16	17,400.88	0.00071
17	17,401.55	0.00065
18	17,401.24	0.00062
19	17,400.78	0.00069
20	17,400.42	0.00072
Average	17,400.90	0.00067

The prototype experimental results of instrument handle trigger in Table 10.17 confirm the appropriate function since the average maximum stress 4,686.52 psi and average maximum deflection 0.00086 in. are approximately same as the results of maximum stress 4,686.99 psi and maximum deflection 0.00097 in. that are laid out, respectively, in Figs. 10.55 and 10.56 by computer-aided modeling and numerical simulation.

Table 10.18 states the prototype testing results of instrument handle in this new biomedical endoscopic surgiclip instrument.

The prototype experimental results of instrument handle in Table 10.18 prove the normal function as the average maximum stress 17,400.90 psi and average maximum deflection 0.00067 in. are close to the results of maximum stress 17,400.49 psi and maximum deflection 0.00074 in. that are shown, respectively in Figs. 10.58 and 10.59 by computer-aided modeling and numerical simulation.

Table 10.19 demonstrates the prototype testing results of instrument front end of inside shaft in this new biomedical endoscopic surgiclip instrument.

The prototype experimental results for instrument front end of inside shaft in Table 10.19 verify the proper function because the average maximum stress 18,835.57 psi and average maximum deflection 0.00113 in. are similar to the results of maximum stress 18,835.02 psi and maximum deflection 0.00106 in. that are

Table 10.19 Prototype testing of instrument front end of inside shaft in this new biomedical endoscopic surgiclip instrument

Number of experiment	Maximum stress (psi)	Maximum deflection (in.)
1	18,835.32	0.00104
2	18,835.12	0.00101
3	18,835.82	0.00135
4	18,834.89	0.00111
5	18,835.88	0.00118
6	18,835.97	0.00124
7	18,835.89	0.00101
8	18,835.54	0.00108
9	18,834.88	0.00112
10	18,835.86	0.00119
11	18,835.88	0.00118
12	18,835.75	0.00119
13	18,835.98	0.00117
14	18,834.78	0.00115
15	18,835.84	0.00101
16	18,835.88	0.00116
17	18,835.95	0.00119
18	18,835.91	0.00118
19	18,835.48	0.00109
20	18,834.75	0.00103
Average	18,835.57	0.00113

presented, respectively, in Figs. 10.61 and 10.62 by computer-aided modeling and numerical simulation.

Table 10.20 demonstrates the prototype testing results of instrument inside shaft in this new biomedical endoscopic surgiclip instrument.

The prototype experimental results of instrument inside shaft in Table 10.20 confirm the normal function since the average maximum stress 20,118.55 psi and average maximum deflection 0.04334 in. are almost equal to the results of maximum stress 20,118.95 psi and maximum deflection 0.04318 in. that are presented, respectively, in Figs. 10.64 and 10.65 by computer-aided modeling and numerical simulation.

Table 10.21 expresses the prototype testing results of instrument front jaw (forming surgiclip to secure the tissues) by using this new biomedical endoscopic surgiclip instrument.

The prototype experimental results of instrument front jaw (forming surgiclip to secure the tissues by using this new biomedical endoscopic surgiclip instrument) in Table 10.21 prove the normal function as the average maximum stress 19,832.61 psi and average maximum deflection 0.00651 in. are very similar to the results of maximum stress 19,832.97 psi and maximum deflection 0.00678 in. that are indicated, respectively, in Figs. 10.67 and 10.68 by computer-aided modeling and numerical simulation.

Table 10.20 Prototype testing of instrument inside shaft in this new biomedical endoscopic surgiclip instrument

Number of experiment	Maximum stress (psi)	Maximum deflection (in.)
1	20,118.85	0.04338
2	20,118.24	0.04324
3	20,118.98	0.04317
4	20,118.25	0.04348
5	20,118.18	0.04338
6	20,118.35	0.04309
7	20,118.38	0.04354
8	20,118.12	0.04378
9	20,118.08	0.04368
10	20,118.11	0.04324
11	20,118.78	0.04338
12	20,118.66	0.04309
13	20,118.54	0.04333
14	20,118.99	0.04338
15	20,118.45	0.04354
16	20,118.84	0.04348
17	20,118.68	0.04329
18	20,118.78	0.04312
19	20,118.99	0.04308
20	20,118.84	0.04319
Average	20,118.55	0.04334

Table 10.21 Prototype testing of instrument front jaw (forming surgiclip to secure the tissues) by using this new biomedical endoscopic surgiclip instrument

Number of experiment	Maximum stress (psi)	Maximum deflection (in.)
1	19,832.88	0.00655
2	19,832.99	0.00688
3	19,832.77	0.00644
4	19,832.54	0.00624
5	19,832.32	0.00618
6	19,832.38	0.00616
7	19,832.18	0.00635
8	19,832.24	0.00638
9	19,832.27	0.00633
10	19,832.77	0.00688
11	19,832.38	0.00668
12	19,832.55	0.00617
13	19,832.48	0.00618
14	19,832.45	0.00666
15	19,832.98	0.00648
16	19,832.54	0.00677
17	19,832.95	0.00679
18	19,832.91	0.00668
19	19,832.78	0.00669
20	19,832.79	0.00675
Average	19,832.61	0.00651

Table 10.22 Prototype testing of instrument front jaw (to manipulate thick tissues) by using this new biomedical endoscopic surgiclip instrument

Number of experiment	Maximum stress (psi)	Maximum deflection (in.)
1	22,916.17	0.00912
2	22,916.24	0.00924
3	22,916.48	0.00901
4	22,916.55	0.00938
5	22,916.68	0.00945
6	22,916.77	0.00955
7	22,916.78	0.00948
8	22,916.44	0.00937
9	22,916.80	0.00976
10	22,916.56	0.00985
11	22,916.88	0.00973
12	22,916.15	0.00908
13	22,916.69	0.00918
14	22,916.54	0.00943
15	22,916.23	0.00901
16	22,916.78	0.00911
17	22,916.54	0.00905
18	22,916.78	0.00924
19	22,916.33	0.00915
20	22,916.52	0.00931
Average	22,916.55	0.00933

Table 10.22 records the prototype testing results of instrument front jaw (to manipulate thick tissues) by using this new biomedical endoscopic surgiclip instrument.

The prototype experimental results of instrument front jaw (to manipulate thick tissues by using this new biomedical endoscopic surgiclip instrument) in Table 10.22 verify the proper function because the average maximum stress 22,916.55 psi and average maximum deflection 0.00933 in. are closely equal to the results of maximum stress 22,916.18 psi and maximum deflection 0.00903 in. that are indicated, respectively, in Figs. 10.69 and 10.70 by computer-aided modeling and numerical simulation.

Table 10.23 shows the prototype testing results of instrument surgiclip pusher by using this new biomedical endoscopic surgiclip instrument.

The prototype experimental results of instrument surgiclip pusher in Table 10.23 confirm the appropriate function since the average maximum stress 20,542.45 psi and average maximum deflection 0.00456 in. are almost same as the results of maximum stress 20,542.72 psi and maximum deflection 0.00429 in. that are displayed, respectively, in Figs. 10.72 and 10.73 by computer-aided modeling and numerical simulation.

Table 10.24 demonstrates the prototype testing results of instrument external shaft in this new biomedical endoscopic surgiclip instrument.

Table 10.23 Prototype testing of instrument surgiclip pusher by using this new biomedical endoscopic surgiclip instrument

Number of experiment	Maximum stress (psi)	Maximum deflection (in.)
1	20,542.62	0.00427
2	20,542.54	0.00435
3	20,542.71	0.00448
4	20,542.52	0.00457
5	20,542.44	0.00478
6	20,542.17	0.00465
7	20,542.18	0.00479
8	20,542.24	0.00488
9	20,542.22	0.00459
10	20,542.38	0.00467
11	20,542.44	0.00478
12	20,542.48	0.00487
13	20,542.66	0.00449
14	20,542.54	0.00438
15	20,542.43	0.00444
16	20,542.38	0.00479
17	20,542.78	0.00432
18	20,542.33	0.00425
19	20,542.45	0.00438
20	20,542.47	0.00451
Average	20,542.45	0.00456

Table 10.24 Prototype testing of instrument external shaft in this new biomedical endoscopic surgiclip instrument

Number of experiment	Maximum stress (psi)	Maximum deflection (in.)
1	12,922.44	0.00078
2	12,922.24	0.00082
3	12,922.18	0.00075
4	12,922.23	0.00071
5	12,922.57	0.00078
6	12,922.31	0.00072
7	12,922.11	0.00071
8	12,922.19	0.00073
9	12,922.59	0.00082
10	12,922.22	0.00073
11	12,922.08	0.00075
12	12,922.17	0.00071
13	12,922.24	0.00072
14	12,922.38	0.00071
15	12,922.27	0.00073
16	12,922.38	0.00071
17	12,922.57	0.00073
18	12,922.12	0.00072
19	12,922.12	0.00071
20	12,922.27	0.00075
Average	12,922.28	0.00074

The prototype experimental results of instrument external shaft in Table 10.24 prove the normal function as the average maximum stress 12,922.28 psi and average maximum deflection 0.00074 in. are almost same as the results of maximum stress 12,922.54 psi and maximum deflection 0.00081 in. that are presented, respectively, in Figs. 10.75 and 10.76 by computer-aided modeling and numerical simulation.

10.4 Discussion and Future Improvement on Biomedical and Surgical Systems

New biomedical open and endoscopic surgiclip instruments have been designed and developed to improve the medical treatment and operation performance during complex medical practices targeting the minimally body-invasive, reliable, and safe medical approaches. The goal of improved biomedical and surgical instruments is to help doctors and surgeons more easily and efficiently handle the patient treatment. The future improvement of biomedical open and endoscopic surgiclip instrument will focus on design simplification and ergonomic satisfaction for consistent function so that doctors can well control and precisely manipulate the medical instruments. Other efforts will be put to modify the front-end mechanism to benefit doctors and surgeons by providing more clear visualization. More clinic field experiment will be planned for further evaluation.

Conclusion

Computer-aided design, 3D modeling, and engineering analysis apply the computer-related technologies and provide scientists, engineers, and technical professionals with research- and engineering-friendly methodologies to efficiently support business product design procedures, engineering documentation, manufacturing drawing, and production processes. The continuous development of computer-aided design, 3D modeling, and engineering analytic technologies will benefit business enterprises, engineering firms, and industrial groups to economically reduce the cost, efficiently shorten the cyclic time, flexibly manipulate the complex 3D geometrical design by computer to avoid unnecessary part making in early system design stages, and effectively perform important engineering analysis to improve designed system quality. The multiple design case studies using computer-aided design applications, 3D modeling, and engineering analysis based on author's research experiences have been introduced in this book. The computer-aided design, numerical simulation, and engineering analytic results included in this book have shown that the computer-aided design and analytic technologies are playing extremely critical fiscal and technological roles in scientific researches, technical renovation, system and product design and development, and industrial and engineering applications. The computer-aided design and engineering analytic technologies bring broad revolution to the business enterprises with its adjustable, flexible, manipulative, reliable, sustainable, and cost-effective control technologies. Multiple case studies of computer-aided design, 3D modeling, and engineering analysis in this book can help to understand and develop more efficient and functional systems to flexibly and convertibly maintain reproducible product quality, shorten required production time, accelerate material-manipulating process, and modify work flow to meet customer needs in their product requirement. Both computer-aided analysis and prototype experiment of different newly designed products that introduced in this book show very close results which validate the credibility and feasibility of fundamental design, 3D modeling, analytic academics, computational simulation, prototyping, and experimental techniques proposed in this book.

J. Zheng Li, *CAD, 3D Modeling, Engineering Analysis, and Prototype Experimentation*, DOI 10.1007/978-3-319-05921-1

Biography

Dr. Jeremy (Zheng) Li is an associate professor of mechanical engineering in the University of Bridgeport. His researches include design and development of robotic systems, automated manufacturing, biomedical and surgical instruments, automated and high-speed machinery, sustainable energy systems (solar panel, wind power, and green energy storage), thermal and flow systems, and structure and stress analysis.

Contact e-mail address: zhengli@bridgeport.edu

J. Zheng Li, *CAD, 3D Modeling, Engineering Analysis, and Prototype Experimentation*, DOI 10.1007/978-3-319-05921-1

References

Abed, H., Atashkari, K., Niazmehr, A., & Jamali, A. (2013). Thermodynamic optimization of combined power and refrigeration cycle using binary organic working fluid. *International Journal of Refrigeration, 36*, 2160–2168.

Adams, S., & Yang, M. (2004). A study on a generalized parametric interpolator with real-time jerk-limited acceleration. *Journal of Computer-Aided Design, 36*, 27–36.

Adán, A., Salamanca, S., & Merchán, P. (2012). A hybrid human computer approach for recovering in complete cultural heritage pieces. *Journal of Computer Graphics, 36*, 1–15.

Agarwal, P., & Manuel, L. (2007). Simulation of offshore wind turbine response for extreme limit states. *Proceedings of 26th International Conference on Offshore Mechanics and Arctic Engineering, USA*.

Agrawal, B., & Karimi, M. (2012). Thermodynamic performance assessment of a novel waste heat based triple effect refrigeration cycle. *International Journal of Refrigeration, 35*, 1647–1656.

Alloway, D. (2000). *Desert survival skills*. Austin, TX: University of Texas Press. ISBN 978-0-292-79226-5.

Ally, M., Munk, J., Baxter, V., & Gehl, A. (2012). Energy analysis and operational efficiency of a horizontal ground-source heat pump system operated in a low-energy test house under simulated occupancy conditions. *International Journal of Refrigeration, 35*, 1092–1103.

American Society of Civil Engineers. (2005). *ASCE 7-05—Minimum design loads for building and other structures*. Reston, VA: ASCE. ISBN 0784408092.

Andrikopoulos, G., Nikolakopoulos, G., Arvanitakis, I., & Manesis, S. (2013). Switching model predictive control of a pneumatic artificial muscle. *International Journal of Control, Automation and Systems, 11*, 1223–1231.

Anjaneyulu, L., Kumar, A., Sankannavar, R., & Rao, K. (2012). Defluoridation of drinking water and rainwater harvesting using a solar still. *Journal of Industrial & Engineering Chemistry Research, 51*, 8040–8048.

Aprea, C., Greco, A., & Maiorino, A. (2013). The substitution of R134a with R744: An exergetic analysis based on experimental data. *International Journal of Refrigeration, 36*, 2148–2159.

Ayou, D., Bruno, J., & Coronas, A. (2012). Steady-state operational degrees of freedom in absorption chillers and heat pumps: Methodology and case study. *International Journal of Refrigeration, 35*, 1570–1582.

Badran, A., Al-Hallaq, A., Salman, I., & Odat, M. (2005). A solar still augmented with a flat-plate collector. *Journal of Desalination, 172*, 227–234.

Bagarella, G., Lazzarin, R., & Lamanna, B. (2013). Cycling losses in refrigeration equipment: An experimental evaluation. *International Journal of Refrigeration, 36*, 2111–2118.

Ballihi, L., BenAmor, B., Daoudi, M., Srivastava, A., & Aboutajdine, D. (2012). Boosting 3D-geometric features for efficient face recognition and gender classification. *IEEE Transactions on Information Forensics and Security, 7*, 1766–1779.

Bang, S., Kim, J., & Lee, J. (2013). An approach of genetic programming for music emotion classification. *International Journal of Control, Automation and Systems, 11*, 1290–1299.

J. Zheng Li, *CAD, 3D Modeling, Engineering Analysis, and Prototype Experimentation*, DOI 10.1007/978-3-319-05921-1

Baniardalani, S., & Askari, J. (2013). Fault diagnosis of timed discrete event systems using dioid algebra. *International Journal of Control, Automation and Systems, 11*, 1095–1105.

Bao, Q., Truong, R., & Lee, B. (2013). Automatic multi-thresholds selection for image segmentation based on evolutionary approach. *International Journal of Control, Automation and Systems, 11*, 834–844.

Barbosa, J., & Sigwalt, R. (2012). Air-side heat transfer and pressure drop in spiral wire-on-tube condensers. *International Journal of Refrigeration, 35*, 939–951.

Bay, H., Ess, A., Tuytelaars, T., & VanGool, L. (2008). Surf: Speed edupro bust features. *Journal of Computer Vision and Image Understanding, 110*, 346–359.

Beebe, R. (2009). Condition monitoring by performance analysis to optimize time for overhaul of centrifugal pumps. *Proceeding of US Annual Meeting for Machinery Failure Prevention Society, 4*, 102–112.

Berretti, S., DelBimbo, A., & Pala, P. (2010). 3D face recognition using iso-geodesic stripes. *IEEE Transactions on Pattern Analysis and Machine Intelligence, 32*, 2162–2177.

Berretti, S., DelBimbo, A., & Pala, P. (2012). Distinguishing facial features for ethnicity-based 3D face recognition. *Proceedings of ACM Transactions on Intelligent Systems and Technology, 3*, 1–20.

Berretti, S., DelBimbo, A., & Pala, P. (2012). Super faces: A super-resolution model for 3D faces. *Proceedings of the Workshop on Non-rigid Shape Analysis and Deformable Image Alignment (NORDIA'12), Firenze, Italy* (pp. 73–82).

Bertoni, M., & Chirumalla, K. (2011). Leveraging web 2.0 in new product development: Lessons learned from a cross-company study. *Journal of Universal Computer Science, 17*, 548–564.

Bhandari, R., & Stadler, I. (2009). Grid parity analysis of solar photovoltaic systems in Germany using experience curves. *Journal of Solar Energy, 83*, 1634–1644.

Bhanja, D., & Kundu, B. (2011). Thermal analysis of a constructal T-shaped porous fin with radiation effects. *International Journal of Refrigeration, 34*, 1483–1496.

Bir, G., & Jonkman, J. (2007). Aeroelastic instabilities of large offshore and onshore wind turbines. *Journal of Physics, 75*, 135–140.

Bouazza, K., & Ouali, M. (2013). Global stabilization of a class of delay discrete-time nonlinear systems via state and output feedback. *International Journal of Control, Automation and Systems, 11*, 1084–1094.

Branker, K., Pathak, M., & Pearce, J. (2011). A review of solar photovoltaic levelized cost of electricity. *Journal of Renewable and Sustainable Energy Reviews, 15*, 4470–4482.

Brinkworth, B., & Sandberg, M. (2006). Design procedure for cooling ducts to minimize efficiency loss due to temperature rise in pv arrays. *Journal of Solar Energy, 80*, 89–103.

Brown, W. T., Wu, X., Wen, B. C., Fowler, J. F., Fayad, F., Amendola, B. E., et al. (2007). Early results of CyberKnife image-guided robotic stereotactic radiosurgery for treatment of lung tumors. *Journal of Computer Aided Surgery, 12*, 253–261.

Carey, V. (2010). Assessment of tesla turbine performance for small scale Rankine combined heat and power systems. *Journal of Engineering for Gas Turbine and Power, 132*, 50–58.

Catalano, C., Mortara, M., Spagnuolo, M., & Falcidieno, B. (2011). Semantics and 3D media: Current issues and perspectives. *Journal of Computer Graphics, 35*, 869–877.

Chae, M. J., Lee, G. W., Kim, J. Y., Park, J. W., & Cho, M. Y. (2011). A 3D surface modeling system for intelligent excavation system. *Journal of Automation in Construction, 20*, 808–817.

Chakraborty, J., Verma, N., & Chhabra, R. P. (2004). Wall effects in the flow past a circular cylinder in a plane channel: A numerical study. *Proceeding of International Conference in Chemical Engineering, 43*, 1529–1537.

Chambers, S., Jarman, J., Etchells, T., & Lisboa, P. (2013). Inference of number of prototypes with a framework approach to K-means clustering. *Journal of Biomedical Engineering and Technology, 13*, 323–340.

Chaouch, M., & Verroust-Blondet, A. (2009). Alignment of 3D models. *Journal of Graph Model, 71*, 63–76.

Che, W., Yang, G., & Jin, X. (2013). Sparse structured non-fragile H∞ controller design for linear systems. *International Journal of Control, Automation and Systems, 11*, 704–710.

Cheng, A., Chiu, P., Chan, P., & Lam, S. (2004). Endoscopic hemostasis for bleeding gastric stromal tumors by application of hemoclip. *Journal of Laparoendoscopic & Advanced Surgical Techniques, 14*, 169–171.

Cho, B., Seo, W., Moon, W., & Baek, K. (2013). Positioning of a mobile robot based on odometry and a new ultrasonic LPS. *International Journal of Control, Automation and Systems, 11*, 333–345.

Cho, C., & Song, J. (2013). Collision detection algorithm robust to model uncertainty. *International Journal of Control, Automation and Systems, 11*, 776–781.

Choi, D., Choi, J., Cho, C., Park, S., & Kang, S. (2013). A safe robot arm with safe joints and gravity compensator. *International Journal of Control, Automation and Systems, 11*, 362–368.

Christian, J., & Hermes, L. (2013). Alternative evaluation of liquid-to-suction heat exchange in the refrigeration cycle. *International Journal of Refrigeration, 36*, 2119–2127.

Christodoulou, F., Giannakakis, P., & Kalfas, A. (2011). Performance benefits of a portable hybrid micro-gas turbine power system for automotive applications. *Journal of Engineering for Gas Turbine and Power, 133*, 86–94.

Christophe, F., Bernard, A., & Coatanéa, E. (2010). A model for knowledge representation of conceptual design. *Journal of Manufacturing Technology, 59*, 155–158.

Chu, F. C., & Chang, B. C. (2005). Automatic visual tracking control system using embedded computers. *Proceeding of the 2005 I.E. International Conference on Mechatronics*, July 10–12.

Claes, P., Smeets, D., Hermans, J., Vandermeulen, D., & Suetens, P. (2011). SHREC'11track: Robust fitting of statistical model. *Proceedings of the Euro Graphics Workshop on 3D Object Retrieval, Llandudno, UK* (pp. 89–95).

Colombo, L., Lucchini, A., & Muzzio, A. (2012). Flow patterns heat transfer and pressure drop for evaporation and condensation of R134A in microfin tubes. *International Journal of Refrigeration, 35*, 2150–2165.

Cong, M., Zhou, Y., Jiang, Y., Kang, R., & Guo, D. (2005). An automated wafer-handling system based on the integrated circuit equipments. *IEEE International Conference on Robotics and Biomimetics, ROBIO, 2*, 240–245.

Coşkun, S., Motorcu, A., Yamankaradeniz, N., & Pulat, E. (2012). Evaluation of control parameters' effects on system performance with Taguchi method in waste heat recovery application using mechanical heat pump. *International Journal of Refrigeration, 35*, 795–809.

Cuevas, C., Fonseca, N., & Lemort, V. (2012). Automotive electric scroll compressor: Testing and modeling. *International Journal of Refrigeration, 35*, 841–849.

Darling, S., You, F., Veselka, T., & Velosa, A. (2011). Assumptions and the levelized cost of energy for photovoltaics. *Journal of Energy Environment, 4*, 3133–3139.

Derking, J., Holland, H., Lerou, P., Tirolien, T., & Brake, H. (2012). Micromachined Joule-Thomson cold stages operating in the temperature range 80–250 K. *International Journal of Refrigeration, 35*, 1200–1207.

Devanathan, S., & Ramani, K. (2010). Creating polytope representations of design spaces for visual exploration using consistency techniques. *Journal of Mechanical Design, 132*, 8–18.

Ding, L., Davies, D., & McMahon, C. (2009). The integration of lightweight representation and annotation for collaborative design representation. *Journal of Research in Engineering Design, 19*, 223–238.

Ding, W., Fan, X., & Liu, X. (2013). Absolute exponential admissibility of switched descriptor delayed systems with sector-bounded nonlinearity. *International Journal of Control, Automation and Systems, 11*, 692–703.

Egolf, G., Vuarnoz, D., Gravier, L., & Courret, G. (2012). The thermodynamics of devices with solid layered-bed magnetocaloric refrigerants. *International Journal of Refrigeration, 35*, 1506–1517.

Ekren, O., Ezan, M., & Erek, A. (2011). Experimental assessment of energy storage via variable speed compressor. *International Journal of Refrigeration, 34*, 1424–1435.

Engelbrecht, K., Eriksen, D., Bahl, C., Bjørk, R., Geyti, J., Lozano, J., et al. (2012). Experimental results for a novel rotary active magnetic regenerator. *International Journal of Refrigeration, 35*, 1498–1505.

Evans, P., Starly, B., & Sun, W. (2006). Computer-aided tissue engineering for design and evaluation of lumbar-spine arthroplasty. *Journal of Computer-Aided Design and Application, 36*, 771–778.

Fadzil, M., Izhar, L., & Nugroho, H. (2011). Analysis of foveal avascular zone for grading of diabetic retinopathy. *Journal of Biomedical Engineering and Technology, 7*, 232–250.

Falconett, I., & Nagasaka, K. (2010). Comparative analysis of support mechanisms for renewable energy technologies using probability distributions. *Journal of Renewable Energy, 35*, 1135–1144.

Faltemier, T. C., Bowyer, K. W., & Flynn, P. J. (2008). A region ensemble for 3D face recognition. *IEEE Transactions on Information Forensics and Security, 3*, 62–73.

Fang, T., Cho, D., & Choi, J. (2013). Optimal scheduling of a communication channel for the centralized control of a platoon of vehicles. *International Journal of Control, Automation and Systems, 11*, 752–760.

Faúndez, C., Barrientos, L., & Valderrama, J. (2013). Modeling and thermodynamic consistency of solubility data of refrigerants in ionic liquids. *International Journal of Refrigeration, 36*, 2242–2250.

Feil-Seifer, D., Skinner, K., & Mataric, M. J. (2007). Benchmarks for evaluating socially assistive robotics. *Journal of Interaction Science, 8*, 423–439.

Ferrey, S. (2006). Renewable orphans: Adopting legal renewable standards at the state level. *Journal of Electricity, 19*, 52–61.

Fuge, M., Yumer, M. E., Orbay, G., & Kara, L. B. (2012). Conceptual design and modification of freeform surfaces using dual shape representations in augmented reality environments. *Journal of Computer-Aided Design, 44*, 1020–1032.

Fulton, G., Malcolm, J., & Moroz, E. (2006). Design of a semi-submersible platform for a 5MW wind turbine. *44th Aerospace Sciences Meeting and Exhibit, USA.*

Fusco, G., & Russo, M. (2013). Design of decentralized robust controller for voltage regulation and stabilization of multimachine power systems. *International Journal of Control, Automation and Systems, 11*, 277–285.

Gao, H., Liu, J., Yu, Y., & Li, Y. (2013). Distance measurement of zooming image for a mobile robot. *International Journal of Control, Automation and Systems, 11*, 782–789.

Gerkey, B., & Matarić, M. J. (2004). A formal framework for the study of task allocation in multi-robot systems. *International Journal of Robotics Research, 23*, 939–954.

Gevorkian, P. (2007). *Solar power in building design (GreenSource): The engineer's complete project resource.* New York, NY: McGraw-Hill Professional. ISBN 0071485635.

Gheisari, R., Jafarian, A., & Ansari, M. (2012). Analytical investigation of compressible oscillating flow in a porous media: A second-order successive approximation technique. *International Journal of Refrigeration, 35*, 1789–1799.

Gill, P., & Munroe, N. (2012). Review on magnesium alloys as biodegradable implant materials. *Journal of Biomedical Engineering and Technology, 10*, 383–398.

Goel, A. K., Vattam, S., Wiltgen, B., & Helms, M. (2012). Cognitive, collaborative, conceptual and creative—Four characteristics of the next generation of knowledge based CAD systems: A study in biologically inspired design. *Journal of Computer-Aided Design, 44*, 879–900.

González-Jorge, H., Liñares, B., Quelle, I., Carballo, E., & Domarco, G. (2007). Measurement of critical current in superconducting rings. *Journal of Measurement, 40*, 378–380.

González-Jorge, H., Peleteiro, J., Carballo, E., & Domarco, G. (2005). Inducting technique and trapped field in ring-shaped superconductors. *Journal of Physics, 433*, 132–134.

Góral, D., & Kluza, F. (2012). Heat transfer coefficient in impingement fluidization freezing of vegetables and its prediction. *International Journal of Refrigeration, 35*, 871–879.

Goy, C., López, M., Bolognini, P., & Herrera, M. (2012). ZigBee-based wireless transmissions interface incorporated to an FPGA embedded system. *Journal of Biomedical Engineering and Technology, 10*, 19–29.

Gupta, S., Markey, M. K., & Bovik, A. C. (2010). Anthropometric 3D face recognition. *International Journal of Computer Vision, 90*, 331–349.

Haddad-Mashadrizeh, A., Matin, M., Bahrami, A., Edalatmanesh, M., Naderi Meshkin, H., Mousavi, S., et al. (2013). Cytotoxicity and biocompatibility evaluation of chitosan-beta glycerol phosphate-hydroxyethyl cellulose hydrogel on adult rat liver for cell-based therapeutic applications. *Journal of Biomedical Engineering and Technology, 13*, 228–239.

Han, G., Guan, Z., Cheng, X., Wu, Y., & Liu, F. (2013). Multiconsensus of second order multiagent systems with directed topologies. *International Journal of Control, Automation and Systems, 11*, 1122–1127.

Han, S., & Lee, J. (2013). Backstepping sliding mode control with FWNN for strict output feedback non-smooth nonlinear dynamic system. *International Journal of Control, Automation and Systems, 11*, 398–409.

Han, C., Wang, W., & Zhang, Y. (2013). White noise estimators for networked systems with packet dropouts. *International Journal of Control, Automation and Systems, 11*, 1187–1195.

Harik, R., Derigent, W., & Ris, G. (2008). Computer aided process planning in aircraft manufacturing. *Journal of Computer-Aided Design and Applications, 5*, 953–962.

Harja, J., Tikkanen, J., Sorvoja, H., & Myllyla, R. (2007). Magnetic resonance imaging compatible three-degrees-of-freedom joystick for surgical robot. *International Journal of Medical Robot, 3*, 365–371.

Hemalatha, K., & Manivannan, M. (2011). A study of cardiopulmonary interaction haemodynamics with detailed lumped parameter model. *Journal of Biomedical Engineering and Technology, 7*, 251–271.

Hirsch, J. (2003). The Lorentz force and superconductivity. *Journal of Physics, 315*, 474–476.

Hoke, A., & Komor, P. (2012). Maximizing the benefits of distributed photovoltaics. *Journal of the Electricity, 25*, 55–67.

Howard, A., Gaurav, S., Sukhatme, S., & Matarić, M. J. (2006). Multi-robot mapping using manifold representations. *Proceedings of the IEEE—Special Issue on Multi-robot Systems, 94*, 1360–1369.

Hu, Z., Chen, Z., Gui, W., Yang, C., Yin, L., & Peng, X. (2013). An efficient multi-PCA based on-line monitoring scheme for multi-stages imperial smelting process. *International Journal of Control, Automation, and Systems, 11*, 317–324.

Hu, H., Yu, L., Chen, G., & Xie, G. (2013). Second-order consensus of multi-agent systems with unknown but bounded disturbance. *International Journal of Control, Automation and Systems, 11*, 258–267.

Isaev, S. A., Baranov, P. A., Kudryavtsev, N. A., Lysenko, D. A., & Usachov, A. E. (2005). Comparative analysis of the calculation data on an unsteady flow around a circular cylinder obtained using the VP2/3 and Fluent packages and the Spalart-Allmaras and Menter turbulence models. *Journal of Engineering Physics and Thermophysics, 78*, 1199–2013.

Jabbar, A., Khalifa, A., & Hamood, M. (2009). Performance correlations for basin type solar stills. *Journal of Desalination, 249*, 24–28.

Janchiv, A., Batsaikhan, D., Kim, B., Lee, W., & Lee, S. (2013). Time-efficient and complete coverage path planning based on flow networks for multi-robots. *International Journal of Control, Automation and Systems, 11*, 369–376.

Janghel, R., Shukla, A., & Tiwari, R. (2012). Hybrid computing based intelligent system for breast cancer diagnosis. *Journal of Biomedical Engineering and Technology, 10*, 1–18.

Jeon, S., Kim, C., Kim, G., Kim, O., & Kee, C. (2013). Optimal signal tracking algorithm for GNSS signal using moving set-point LQG system. *International Journal of Control, Automation and Systems, 11*, 1214–1222.

Jeong, K., Choi, J., & Lee, B. (2013). Reliability-based camera handoff for cooperative tracking with multiple pan-tilt cameras. *International Journal of Control, Automation and Systems, 11*, 815–825.

Ji, Y., Zong, Q., Tian, B., & Liu, H. (2013). Input-to-state-stability modular command filtered back-stepping attitude control of a generic reentry vehicle. *International Journal of Control, Automation and Systems, 11*, 734–741.

Jiang, C. (2013). Observer design for neutral systems with delays and nonlinear perturbations: Delay-dependent design method. *International Journal of Control, Automation and Systems, 11*, 427–432.

Kagawa, N., Matsuguchi, A., Yamaya, K., & Watanabe, K. (2013). Behavior of isobaric heat capacity of R32 in the gas phase. *International Journal of Refrigeration, 36*, 2216–2222.

Kalantzaki, K., Bei, K., Garofalakis, M., & Zervakis, M. (2013). Biological interaction networks based on non-parametric estimation. *Journal of Biomedical Engineering and Technology, 13*, 383–409.

Kavitha, G., & Ramakrishnan, S. (2011). Detection of blood vessels in human retinal images using Ant Colony Optimisation method. *Journal of Biomedical Engineering and Technology, 5*, 360–370.

Kayalvizhi, M., Kavitha, G., & Sujatha, C. (2013). Analysis of ventricle regions in Alzheimer's brain MR images using level set based methods. *Journal of Biomedical Engineering and Technology, 13*, 300–319.

Kazerooni, H. (2005). Exoskeletons for human power augmentation. *Proceeding of IEEE Conference of Intelligent Robots and Systems, Edmonton, Canada* (pp. 3459–3464).

Kennedy C. W., & Desai J. P. (2003). Force feedback using vision. *The 11th International Conference on Advanced Robotics*. Portugal: University of Coimbra.

Khan, M., Khan, T., Chyu, M., & Ayub, Z. (2012). Experimental investigation of evaporation heat transfer and pressure drop of ammonia in a 30° chevron plate heat exchanger. *International Journal of Refrigeration, 35*, 1757–1765.

Kim, K. Y., & Kim, Y. S. (2011). Causal design knowledge: Alternative representation method for product development knowledge management. *Journal of Computer Aided Design, 43*, 1137–1153.

Kim, D., Kim, D., Choi, W., & Kang, J. (2011). Dynamics simulations of a graphite block under longitudinal impact. *Journal of Engineering for Gas Turbine and Power, 133*, 38–45.

Kim, E., Kwak, K., Kim, Y., Kim, S., Kwak, B., Jang, I., et al. (2013). Auto-positioning of sliding planes based on virtual force. *International Journal of Control, Automation, and Systems, 11*, 798–804.

Kim, T., Moon, K., & Song, T. (2013). Variable-structured interacting multiple model algorithm for the ballistic coefficient estimation of a re-entry ballistic target. *International Journal of Control, Automation and Systems, 11*, 1204–1213.

Kitanovski, A., Egolf, P., & Poredos, A. (2012). Rotary magnetic chillers with permanent magnets. *International Journal of Refrigeration, 35*, 1055–1066.

Klopcar, N., Tomsic, M., & Lenarcic, J. (2007). A kinematic model of the shoulder complex to evaluate the arm-reachable workspace. *Journal of Biomechanics, 40*, 86–91.

Komandur, S., & Sunder, R. (2008). Last stage performance considerations in low-pressure turbines of power plants: A case study. *Journal of Engineering for Gas Turbine and Power, 130*, 38–45.

Kong, K., & Tomizuka, M. (2013). Real-time nonlinear programming by amplitude modulation. *International Journal of Control, Automation and Systems, 11*, 742–751.

Koo, G., Park, J., & Joo, Y. (2013). Intelligent digital redesign for nonlinear systems using a guaranteed cost control method. *International Journal of Control, Automation and Systems, 11*, 1075–1083.

Koo, I., Tran, D., Lee, Y., Moon, H., Ja Koo, C., Park, S., et al. (2013). Development of a quadruped walking robot AiDIN-III using biologically inspired kinematic analysis. *International Journal of Control, Automation and Systems, 11*, 1276–1289.

Kosmadoudia, Z., Lima, T., Ritchie, J., Louchart, S., Liu, Y., & Sunga, R. (2013). Engineering design using game-enhanced CAD: The potential to augment the user experience with game elements. *Journal of Computer Aided Design, 45*, 777–795.

Kosta, S., Kosta, Y., Chaudhary, J., Vaghela, P., Mehta, H., Patel, B., et al. (2012). Bio-material human body part (palm fingers) based electronic FET transistor. *Journal of Biomedical Engineering and Technology, 10*, 368–382.

Kumar, S., Kumar, A., Gombar, S., Trikha, A., & Anand, S. (2012). Fractal dimension of electroencephalogram for assessment of hypnosis state of patient during anaesthesia. *Journal of Biomedical Engineering and Technology, 10*, 30–37.

Kumlutaş, D., Karadeniz, Z., & Özşen, H. (2012). Investigation of design parameters of a domestic refrigerator by artificial neural networks and numerical simulations. *International Journal of Refrigeration, 35*, 1678–1689.

Kundu, P. K., & Cohen, I. M. (2008). *Fluid mechanics*. Burlington, MA: Academic Press. ISBN 978-0-123-73735-9.

Laird, J. (2011). PV's falling costs: In the U.S., the DOE is pioneering research in order to reduce the cost of installed PV to below a dollar-per-Watt by 2017. *Journal of Renewable Energy Focus, 12*, 52–56.

Lamooki, G. (2013). Recursive partial stabilization: Backstepping and generalized strict feedback form. *International Journal of Control, Automation and Systems, 11*, 250–257.

Lattemann, S., & Höpner, T. (2008). Environmental impact and impact assessment of seawater desalination. *Journal of Desalination, 220*, 1–15.

Laufer, L., Anand, V., & Schwartz, T. (2007). Endoscopic, endonasal extended transsphenoidal, transplanum transtuberculum approach for resection of suprasellar lesions. *Journal of Neurosurgery, 106*, 400–406.

Lee, S., & Jung, S. (2013). Object handling control among two-wheel robots and a human operator: An empirical approach. *International Journal of Control, Automation and Systems, 11*, 346–353.

Lee, H., Kim, J., & Banerjee, A. (2010). Collaborative intelligent CAD framework incorporating design history tracking algorithm. *Journal of Computer-Aided Design, 42*, 1125–1142.

Lee, H., Lee, B., Jung, J., Hong, B., & Kim, E. (2013). Human biometric identification through integration of footprint and gait. *International Journal of Control, Automation and Systems, 11*, 826–833.

Lee, S., Ma, Y. S., Thimm, G., & Verstraeten, J. (2008). Product lifecycle management in aviation maintenance, repair and overhaul. *Journal of Computers in Industry, 59*, 296–303.

Leibe, B., Leonardis, A., & Schiele, B. (2008). Robust object detection with interleaved categorization and segmentation. *International Journal of Computer Vision, 77*, 259–289.

Lerman, K., Jones, C., Galstyan, A., & Mataric, M. J. (2006). Analysis of dynamic task allocation in multi-robot systems. *International Journal of Robotics Research, 25*, 225–242.

Li, J. Z. (2009d). Study and development of an energy-saving mechanical system. *International Journal of Recent Trends in Engineering, 1*, 51–54.

Li, J. Z. (2011a). Design and development of a new automated and high speed gas filling systems. *Journal of Mechanical Engineering (International Scholarly Research Network), 1*, 1–4.

Li, J. Z. (2011b). A new automated and high speed machinery system for high viscous liquids. *Journal of Applied Mechanics and Materials, 66–68*, 140–145.

Li, J. Z. (2011c). Computer-aided design, modeling and simulation of a new solar still system design. *Journal of Modeling and Simulation in Engineering, 2*, 1–5.

Li, J. Z. (2012d). Study of a new biomedical and endoscopic surgical instrument. *American Journal of Biomedical Engineering, 2*, 12–22.

Li, J. Z. (2012f). Study and analysis of turbine blade design in a new wind power system. *Journal of Mechanical Engineering (i-Manager's Research Network), 2*(3 May-July), 1–4.

Li, H., Huang, D., Lemaire, P., Morvan, J. M., & Chen, L. (2011). Expression robust 3D face recognition via mesh-based histograms of multiple order surface differential quantities.

Proceedings of the IEEE International Conference on Image Processing (ICIP'11) (pp. 3053–3056).

Li, Y., & Jian, J. (2010). Development of key technologies in a case-based knowledge system for fixture design. *Journal of Digital Enterprise Technology, 66*, 239–250.

Li, J., Kim, B. C., & Han, S. (2012). Parametric exchange of round shapes between a mechanical CAD system and a ship CAD system. *Journal of Computer-Aided Design, 44*, 154–161.

Li, D., Xu, H., He, X., & Lan, H. (2005). Theoretical and experimental study on the magnetic fluid seal of reciprocating shaft. *Journal of Magnetism and Magnetic Materials, 2*, 399–402.

Liaquat, M., & Malik, M. (2013). On sampled data regulation of feedback linearizable systems using a realizable reconstruction filter. *International Journal of Control, Automation and Systems, 11*, 1162–1169.

Lin, K., Kuo, C., Hsieh, W., & Wang, C. (2013). Modeling and simulation of the transcritical CO2 heat pump system. *International Journal of Refrigeration, 36*, 2048–2064.

Lin, H., Lo, W., Cheng, Y., & Peng, C. (2007). Endoscopic hemoclip versus triclip placement in patients with high risk peptic ulcer bleeding. *Journal of Gastroenterol, 102*, 539–543.

Lini, G., Piazzi, A., & Consolini, L. (2013). Algebraic solution to minimum-time velocity planning. *International Journal of Control, Automation and Systems, 11*, 805–814.

López-Sastre, R. J., García-Fuertes, N. A., Redondo-Cabrera, C., Acevedo-Rodríguez, F. J., & Maldonado-Bascón, J. (2013). Evaluating 3D spatial pyramids for classifying 3D shapes. *Journal of Computers & Graphics, 37*, 473–483.

Lucas, C., & Koehler, K. (2012). Experimental investigation of the COP improvement of a refrigeration cycle by use of an ejector. *International Journal of Refrigeration, 35*, 1595–1603.

MacLeod, J., & Jastremski, J. (2010). Development of a unique icing spray system for a new facility for certification of large turbofan engines. *SAE International Journal of Aerospace, 2*, 238–248.

Manikandan, V., Shanmugasundaram, M., Shanmugan, S., Janarthanan, M., & Chandrasekaran, J. (2013). Wick type solar stills: A review. *Journal of Renewable and Sustainable Energy Reviews, 20*, 322–335.

Mehta, K. C., & Coulburne, W. L. (2010). *Wind loads: Guide to the wind load provisions of ASCE 7-05*. Reston, VA: ASCE Press. ISBN 978-0-7844-0858-2.

Mellinger, D., Michael, N., & Kumar, V. (2010). Trajectory generation and control for precise aggressive maneuvers with quadrotors. *Proceedings of International Symposium on Experimental Robotics, Delhi, India.*

Mendonça, M., & Jacobs, D. (2009). Feed-in tariffs go global: Policy in practice. *World Magazine of Renewable Energy, 12*, 38–48.

Mian, A. S., Bennamoun, M., & Owens, R. (2008). Key point detection and local feature matching for textured 3D face recognition. *International Journal of Computer Vision, 79*, 1–12.

Mian, A. S., Bennamoun, M., & Owens, R. (2010). On the repeatability and quality of key points for local feature-based 3D object retrieval from cluttered scenes. *International Journal of Computer Vision, 89*, 348–361.

Mitamura, Y., Arioka, S., Sakota, D., Sekine, K., & Azegami, M. (2008). Application of a magnetic fluid seal to rotary blood pumps. *Journal of Physics: Condensed Matter, 2*, 204–208.

Morton, G., & Fruh, G. (2002). The force on an object passing through a magnetic fluid seal. *Journal of Magnetism and Magnetic Materials, 2*, 324–326.

Mumanachit, P., Reindl, D., & Nellis, G. (2012). Comparative analysis of low temperature industrial refrigeration systems. *International Journal of Refrigeration, 35*, 1208–1221.

Munilla, R. (2013). Practical survivor retrieved. *Journal of Solar Energy*, 38–42.

Myers, K., Klein, S., & Reindl, D. (2010). Assessment of high penetration of solar photovoltaics in Wisconsin. *Journal of Energy Policy, 38*, 7338–7345.

Nikolakopoulos, G., & Alexis, K. (2013). Switching networked attitude control of an unmanned quadrotor. *International Journal of Control, Automation and Systems, 11*, 389–397.

Ochonski, W. (2005). New designs of magnetic fluid exclusion seals for rolling bearings. *Journal of Industrial Lubrication and Tribology, 5*, 107–115.

Ogbonnaya, E. (2011). Gas turbine performance optimization using compressor online water washing technique. *Journal of Engineering, 3*, 500–507.

Ogbonnaya, E., Ugwu, H., & Johnson, C. (2010). Computer-aided solution to the vibrational effect of instabilities in gas turbine compressors. *Journal of Engineering, 2*, 658–664.

Oró, E., Miró, L., Farid, M., & Cabeza, L. (2012a). Improving thermal performance of freezers using phase change materials. *International Journal of Refrigeration, 35*, 984–991.

Oró, E., Miró, L., Farid, M., & Cabeza, L. (2012b). Thermal analysis of a low temperature storage unit using phase change materials without refrigeration system. *International Journal of Refrigeration, 35*, 1709–1714.

Ouertani, M., Baina, S., Gzara, L., & Morel, G. (2011). Traceability and management of dispersed product knowledge during design and manufacturing. *Journal of Computer Aided Design, 43*, 546–562.

Pai, M. (2013). Observer-based adaptive sliding mode control for robust tracking and model following. *International Journal of Control, Automation and Systems, 11*, 225–232.

Park, W., Kim, D., & Lee, H. (2013). Terrain trafficability analysis for autonomous navigation: A GIS-based approach. *International Journal of Control, Automation and Systems, 11*, 354–361.

Park, H., Lee, J., Kim, W., & Kim, W. (2012). Performance optimization of a hybrid ground source heat pump with the parallel configuration of a ground heat exchanger and a supplemental heat rejecter in the cooling mode. *International Journal of Refrigeration, 35*, 1537–1546.

Passalis, G., Perakis, P., Theoharis, T., & Kakadiaris, A. (2011). Using facial symmetry to handle pose variation sin real-world 3D face recognition. *IEEE Transactions on Pattern Analysis and Machine Intelligence, 33*, 1938–1951.

Passon, P., Kühn, M., Butterfield, S., & Jonkman, B. (2007). OC3-benchmark exercise of aero-elastic offshore wind turbine codes. *Journal of Physics, 78*, 120–124.

Pessoa, S. A., deS, A., Moura, G., Lima, J. P., Teichrieb, V., & Kelner, J. (2012). Real-time photo realistic rendering of synthetic objects into real scenes. *Journal of Computer Graphics, 36*, 50–69.

Piacentino, A., & Talamo, M. (2013). Innovative thermoeconomic diagnosis of multiple faults in air conditioning units: Methodological improvements and increased reliability of results. *International Journal of Refrigeration, 36*, 2343–2365.

Piatt, J., Starly, B., Faerber, E., & Sun, W. (2006). Application of computer-aided design methods in craniofacial reconstructive surgery using a commercial image-guidance system. *Journal of Neurosurgery, 104*, 64–67.

Polat, K., & Kırmacı, V. (2011b). Determining of gas type in counter flow vortex tube using pairwise fisher score attribute reduction method. *International Journal of Refrigeration, 34*, 1372–1386.

Pounds, P., Mahony, R., Gresham, J., Corke, P., & Roberts, J. (2004). Towards dynamically-favourable quad-rotor aerial robots. *Australasian Conference on Robotics and Automation*. Canberra, Australia.

Qureshi, B., & Zubair, S. (2013). Mechanical sub-cooling vapor compression systems: Current status and future directions. *International Journal of Refrigeration, 36*, 2097–2110.

Rachidi, S., Channa, R., & Karama, A. (2013). 2D inverse problem in a distributed parameter system. *International Journal of Control, Automation and Systems, 11*, 1106–1111.

Ramadhan, M., & Naseeb, A. (2011). The cost benefit analysis of implementing photovoltaic solar system in the state of Kuwait. *Journal of Renewable Energy, 36*, 1272–1276.

Ratti, J., & Vachtsevanos, G. (2010). A biologically-inspired micro aerial vehicle sensing, modeling and control strategies. *Journal of Intelligent & Robotic Systems, 60*, 153–178.

Regli, W. C., Kopena, J. B., & Grauer, M. (2011). On the long term retention of geometry-centric digital engineering artifacts. *Journal of Computer Aided Design, 43*, 820–837.

Reich, Y., & Paz, A. (2008). Managing product quality, risk, and resources through resource quality function deployment. *Journal of Engineering Design, 19*, 249–267.

Rezgui, Y., Boddy, S., Wetherill, M., & Cooper, G. (2011). Past, present and future of information and knowledge sharing in the construction industry: Towards semantic service-based e-construction. *Journal of Computer Aided Design, 43*, 502–515.

Rocca, G. L. (2012). Knowledge based engineering: Between ai and cad review of a language based technology to support engineering design. *Journal of Advanced Engineering Informatics, 26*, 159–179.

Salti, S., Tombari, F., & DiStefano, L. (2013). Performance evaluation of 3D key point detectors. *International Journal of Computer Vision, 102*, 198–220.

Sami, M., & Patton, R. (2013). Active fault tolerant control for nonlinear systems with simultaneous actuator and sensor faults. *International Journal of Control, Automation and Systems, 11*, 1149–1161.

Sanaye, S., & Asgari, H. (2013). Thermal modeling of gas engine driven air to water heat pump systems in heating mode using genetic algorithm and Artificial Neural Network methods. *International Journal of Refrigeration, 36*, 2262–2277.

Saravanamuttoo, H., Rogers, G., & Cohen, H. (2009). *Gas turbine theory*. New York, NY: Prentice Hall. ISBN 10: 0132224372.

Šarevski, V., & Šarevski, M. (2012). Energy efficiency of the thermocompression refrigerating and heat pump systems. *International Journal of Refrigeration, 35*, 1067–1079.

Seifabadi, R., Rezaei, S., Ghidary, S., & Zareinejad, M. (2013). A teleoperation system for micro positioning with haptic feedback. *International Journal of Control, Automation and Systems, 11*, 768–775.

Senthil, K., Chandrasegaran, K., Ramania, R. D., Imré, H., Alain, B., Ramy, F., et al. (2013). The evolution, challenges, and future of knowledge representation in product design systems. *Journal of Computer Aided Design, 45*, 204–228.

Seok, J., Lee, J., Wang, J., Lee, J., & Lee, H. (2013). A temporal path planner for solving information inconsistency in an integrated path planner. *International Journal of Control, Automation and Systems, 11*, 1232–1240.

Shakernia, O., Vidal, R., Sharp, C. S., Ma, Y., & Sastry, S. (2002) Multiple view motion estimation and control for landing an unmanned aerial vehicle. *Proceedings of IEEE International Conference on Robotics and Automation* (pp. 2793–2798).

Shamsoddini, R., & Khorasani, A. (2012). A new approach to study and optimize cooling performance of a Ranque–Hilsch vortex tube. *International Journal of Refrigeration, 35*, 2339–2348.

Shen, C., He, Y., Liu, Y., & Tao, W. (2008). Modelling and simulation of solar radiation data processing with Simulink. *Journal of Simulation Modeling Practice and Theory, 2*, 721–735.

Shlyk, L., Krabbes, G., Fuchs, G., Nenkov, K., & Schupp, B. (2004). Flux pinning and magnetic relaxation in melt-processed $YBa_2Cu_3O_{7-d}$ doped with Li. *Journal of Applied Physics, 96*, 371–374.

Silva, D., Barbosa, J., & Adade, A. (2011). Multivariable aircraft engine controller design using an optimal loop shaping approach. *SAE International Journal of Aerospace, 4*, 661–671.

Simhauser, P. (2010). The hidden costs of wind generation in a thermal power system: What cost? *Journal of Australian Economic Review, 44*, 269–292.

Singh, P., & Nestmann, F. (2011). A consolidated model for the turbine operation of centrifugal pumps. *Journal of Engineering for Gas Turbine and Power, 133*, 102–111.

Sipiran, I., & Bustos, B. (2010). Arobust 3D interest points detector based on Harris operator. *Proceedings of the Euro Graphics Workshop on 3D Object Retrieval* (pp. 7–14). Norrköping, Sweden: Euro Graphics Association.

Song, Z., & Li, H. (2013). Second-order sliding mode control with backstepping for aeroelastic systems based on finite-time technique. *International Journal of Control, Automation and Systems, 11*, 416–421.

Starly, B., Fang, Z., Sun, W., & Regli, W. (2005). Three-dimensional reconstruction for medical-CAD modeling. *Journal of Computer-Aided Design and Application, 2*, 431–438.

Stefano, B., Naoufel, W., Albertodel, B., & Pietro, P. (2013). Matching 3D face scans using interest points and local histogram descriptors. *Journal of Computers & Graphics, 37*, 509–525.

Sun, W., Starly, B., Nam, J., & Darling, A. (2005). Bio-CAD modeling and its application in computer-aided tissue engineering. *Journal of Computer- Aided Design, 37*, 1097–1114.

Sung, R., Ritchie, J. M., Rea, H. J., & Corney, J. (2011). Automated design knowledge capture and representation in single-user cad environments. *Journal of Engineering Design, 22*, 487–503.

Takashi, K. (2002). Vacuum manipulator for semiconductor manufacturing equipment. *Journal of Industrial Robot, 2*, 324–328.

Tapus, A., Cristian, T., & Mataric, M. J. (2008). User-robot personality matching and assistive robot behavior adaptation for post-stroke rehabilitation therapy. *Journal of Intelligent Robotics, 5*, 169–183.

Tapus, A., Matarić, M. J., & Scassellati, B. (2007). The grand challenges in socially assistive robotics. *IEEE Robotics and Automation Magazine, 14*, 15–18.

Thomas, R., Ghosh, P., & Chowdhury, K. (2012). Exergy based analysis on different expander arrangements in helium liquefiers. *International Journal of Refrigeration, 35*, 1188–1199.

Tian, C., Masry, M., & Lipson, L. (2009). Physical sketching: Reconstruction and analysis of 3D objects from freehand sketches. *Journal of Computer Aided Design, 41*, 147–158.

Timilsina, G., Kurdgelashvili, L., & Narbel, P. (2012). Solar energy: Markets, economics and policies. *Journal of Renewable and Sustainable Energy Reviews, 16*, 449–465.

Tiwari, A., & Tiwari, G. (2007). Thermal modeling based on solar fraction and experimental study of the annual and seasonal performance of a single slope passive solar still: The effect of water depths. *Journal of Desalination, 207*, 184–204.

Tombari, F., Salti, S., & DiStefano, L. (2010). Unique shape context for 3D data description. *Proceedings of the ACM Workshop on 3D Object Retrieval, Firenze, Italy* (pp. 57–62).

Tombari, F., Salti, S., & DiStefano, L. (2010). Unique sig nature of his to grams for local surface description. *European Conference on Computer Vision, Heraklion, Crete, Greece, 3*, 347–360.

Topal, U., Dorosiskii, L., Ozkan, H., & Yabuz, H. (2003). Effect of thermal neutron irradiation in boron-doped melt-textured YBCO superconductors. *Journal of Physics, 388*, 401–403.

Torbjorn, S. D., Mataric, M. J., & Gaurav, S. S. (2009). Multi-robot task allocation through vacancy chain scheduling. *Journal of Robotics and Autonomous Systems, 3*, 51–57.

Tsagarakis, N., & Caldwell, D. (2003). Development and control of a "soft-actuated" exoskeleton for use in physiotherapy and training. *Journal of Autonomous Robots, 15*, 21–33.

Tuan, L., Lee, S., Dang, V., Moon, S., & Kim, B. (2013). Partial feedback linearization control of a three-dimensional overhead crane. *International Journal of Control, Automation and Systems, 11*, 718–727.

Tušek, J., Kitanovski, A., Prebil, I., & Poredoš, A. (2011). Dynamic operation of an active magnetic regenerator (AMR): Numerical optimization of a packed-bed AMR. *International Journal of Refrigeration, 34*, 1507–1517.

Ullah, I., Ullah, F., Ullah, Q., & Shin, S. (2013). Integrated tracking and accident avoidance system for mobile robots. *International Journal of Control, Automation and Systems, 11*, 1253–1265.

Vallee, C., Beyer, M., Jucas, D., & Carl, H. (2009). Air/water counter-current flow experiments in a model of the hot leg of a pressurized water reactor. *Journal of Engineering for Gas Turbine and Power, 131*, 38–46.

Van, M., Kang, H., Suh, Y., & Shin, K. (2013). A robust fault diagnosis and accommodation scheme for robot manipulators. *International Journal of Control, Automation and Systems, 11*, 377–388.

Veltkamp, R., vanJole, S., Drira, H., BenAmor, B., Daoudi, M., & Li, H., et al. (2011). SHREC'11 track: 3D face models retrieval. *Proceedings of the Euro Graphics Workshop on 3D Object Retrieval. Llandudno, UK* (pp. 89–95).

Vincent, A., Cicirello, A., William, C., & Regli, B. (2013). A flexible and extensible approach to automated CAD/CAM format classification. *Journal of Computers & Graphics, 37*, 484–495.

Walthall, C., Devanathan, S., Ramani, K., Hirleman, E., Kisselburgh, L., & Yang, M. (2011). Evaluating wikis as a communicative medium for collaboration within colocated and distributed design teams. *Journal of Mechanical Design, 133*, 71–81.

Wang, D., Ding, F., & Zhu, D. (2013). Data filtering based least squares algorithms for multivariable CARAR-like systems. *International Journal of Control, Automation and Systems, 11*, 711–717.

Wang, X., Kurdgelashvili, L., Byrne, J., & Barnett, A. (2011). The value of module efficiency in lowering the levelized cost of energy of photovoltaic systems. *Journal of Renewable and Sustainable Energy Reviews, 15*, 4248–4254.

Wang, W., Ma, S., & Zhang, C. (2013). Stability and static output feedback stabilization for a class of nonlinear discrete-time singular switched systems. *International Journal of Control, Automation and Systems, 11*, 1138–1148.

Wang, H., & Yang, G. (2013). Robust filter design for affine fuzzy systems. *International Journal of Control, Automation and Systems, 11*, 410–415.

White, F. M. (2003). *Fluid mechanics*. Boston, MA: McGraw-Hill. ISBN 0072402172.

Yang, D., & Kim, D. (2013). A fast and efficient image registration algorithm using outlier rejection technique based on subimage. *International Journal of Control, Automation and Systems, 11*, 790–797.

Yang, T., Ma, Z., & Yang, Q. (2011). Formation and performance of Kaolin/MnO_2 bi-layer composite dynamic membrane for oily wastewater treatment: Effect of solution conditions. *Journal of Desalination, 270*, 50–56.

Yoon, M. (2013). Single agent control for cyclic consensus systems. *International Journal of Control, Automation and Systems, 11*, 243–249.

Zhao, M., Zou, J., & Hu, J. (2006). An analysis on the magnetic fluid seal capacity. *Journal of Magnetism and Magnetic Materials, 3*, 428–431.

Zhu, Q. (2013). Stabilization of stochastically singular nonlinear jump systems with unknown parameters and continuously distributed delays. *International Journal of Control, Automation and Systems, 11*, 683–691.

Zhu, Y., Jin, X., Du, Z., Fan, B., & Fu, S. (2013). Generic simulation model of multi-evaporator variable refrigerant flow air conditioning system for control analysis. *International Journal of Refrigeration, 36*, 1602–1615.

Zydlo, Z., Ochonski, W., & Zachara, B. (2005). Experiments on magnetic fluid rotary seals operating under vacuum conditions. *Joural of TriboTest, 11*, 345–354.

Index

J. Zheng Li, *CAD, 3D Modeling, Engineering Analysis, and Prototype Experimentation*, DOI 10.1007/978-3-319-05921-1

W

CPSIA information can be obtained
at www.ICGtesting.com
Printed in the USA
LVHW080120070519
616813LV00002B/28/P

9 783319 059204